Public Procurement of Energy Efficiency Services

Public Procurement of Energy Efficiency Services

Lessons from International Experience

Jas Singh
Dilip R. Limaye
Brian Henderson
Xiaoyu Shi

THE WORLD BANK
Washington, DC

ESMAP
Energy Sector Management Assistance Program

1818 H Street NW
Washington DC 20433
Telephone: 202-473-1000
Internet: www.worldbank.org
E-mail: feedback@worldbank.org

ISBN-13: 978-0-8213-8062-8
eISBN: 978-0-8213-8102-1
DOI: 10.1596/978-0-8213-8062-8

Cover photo: Stan Constantio, The World Bank
Cover design: Quantum Think

Library of Congress Cataloging-in-Publication Data

Public procurement of energy efficiency services : lessons from international experience / Jas Singh ... [et al.].
    p. cm.
  Includes bibliographical references and index.
  ISBN 978-0-8213-8062-8 — ISBN 978-0-8213-8102-1 (electronic)
  1. Public contracts—Case studies. 2. Performance contracts—Case studies. 3. Energy conservation—Case studies. I. Singh, Jas, 1966-
  HD3860.P78 2009
  352.5'6—dc22
                                                                    2009027970

Contents

Boxes

Figures

Tables

Foreword

For the past two decades, the World Bank Group (WBG) has been engaged in promoting energy efficiency. At the 2004 Bonn International Conference for Renewable Energies, the WBG committed to increase financing for renewable energy and energy efficiency operations by 20 percent per year over the next five years. Since then, investment operations for energy efficiency have grown steadily, from $177 million in fiscal 2003 to nearly $1.7 billion in fiscal 2009. These projects have addressed the full range of end use and supply-side opportunities and have focused on removing institutional, regulatory, financial, and technical barriers. The WBG's commitment to energy efficiency is further reinforced through its key role in leading the global cooperative effort to reduce greenhouse gas emissions through the Clean Energy Investment Framework and subsequent Strategic Framework on Climate Change and Development.

Energy efficiency remains as important as ever to the WBG and our client countries, in view of universal concerns over global energy security, competitiveness, and environmental protection. While energy efficiency can alleviate pressures in all three areas, realizing large-scale energy savings is a significant challenge for our client countries. Questions persist on how best to identify, package, and finance many small, dispersed projects in a given market. Other informational, technical, financial, and behavioral barriers remain, thwarting efforts to convince end users to reduce their energy waste. Whereas some promising models from the developed world exist, difficulties lie in adapting them to fit the conditions and markets in the developing world.

This book looks at a largely untapped energy efficiency market—the public sector. The potential for efficiency gains in the public sector is substantial, but the implementation of energy savings programs is complicated by numerous factors, including a lack of commercial orientation on the part of public agencies, limited incentives to lower energy costs, complex and strict budgeting and procurement procedures, and limited access to budgetary or project financing. Many public agencies, particularly in developing countries, face severe budget constraints and often focus on upfront costs as a matter of necessity. That leads to greater operating cost liabilities in out-years, causing further budgetary pressures and creating a vicious cycle. Surmounting these barriers is a challenge in both developed and developing countries.

This book explores energy savings performance contracts (ESPCs) as a means of overcoming some of the more difficult hurdles in promoting energy efficiency in public facilities. ESPCs represent a very attractive solution to many of the problems that are unique to public agencies, since they involve outsourcing a full project cycle to a service provider. From the detailed audit through implementation and savings verification, ESPCs can relieve public agencies of bureaucratic hassles, while service providers can secure the off-budget project financing and be paid from the actual energy savings, thus internalizing project performance risks. ESPC bidding also allows public agencies to select from a range of technical solutions, maximizing the benefit to the agency. Global experience suggests that ESPCs have been more effective at realizing efficiency gains than many other policy measures and programs, since the service providers have a vested interest in ensuring that a project is actually implemented. Many of the country governments interviewed for the study also saw enormous potential in bundling, financing, and implementing energy efficiency projects on a larger scale in the public sector, a method that increases the rate of efficiency gains and creates further benefits through economies of scale.

The program and project case studies cited in this book identify approaches, models, and specific solutions to some of the more vexing problems in public procurement for ESPCs, ranging from budgeting, to conducting energy audits, to bid evaluation. The country governments surveyed point out many budgeting, procurement, and other challenges with the use of ESPCs, but they also acknowledge promising approaches and emerging models in this area. What is perhaps most exciting is the potential to replicate this on a scale previously unseen. If governments are able to package 50 to 100 (or more) similar public facilities at a time, and bid out their renovation and retrofit to commercial energy service providers, while mobilizing commercial financing, significant energy savings could be achieved.

The book recommends a demand-driven and flexible process, in which specific procurement provisions and procedures are tailored to a country or public agency, based on local regulations and conditions and on client needs. Equally important is that such tenders could encourage a wide variety of enterprises—from equipment suppliers, to engineering firms, to full-service energy service companies—to use their expertise to develop financially attractive energy efficiency projects for the public client. The public client could then select a proposal that offers the best value and package of services.

Such an approach can also serve as an attractive element in fiscal stimulus packages and efforts by governments to "green" their infrastructure. Retrofits of large bundles of facilities can be a way to disburse funds rapidly, while upgrading facilities, creating local jobs, reducing future operating costs, and mitigating the carbon footprint of public properties. Lower energy bills, in turn, help to create fiscal space in future years for governments to expand social services and meet other critical investment priorities. Governments can thus help stimulate the local market for energy efficiency goods and services and lead by example, demonstrating good practices in energy management and providing models to the private sector to realize such opportunities.

Jamal Saghir
Director, Energy, Transport, and Water
Chair, Energy Sector Board
Sustainable Development Network
The World Bank

Acknowledgments

This book presents the results of the international study, "Public Procurement of Energy Efficiency Services" (P112187), which was undertaken and financed by the Energy Sector Management Assistance Program (ESMAP) in the Energy, Transport and Water Department (ETW) of the World Bank in 2008–09.

The task team included Jas Singh (Task Team Leader and lead author), Xiaoyu Shi (ESMAP Operations Analyst), Dilip R. Limaye, Brian Henderson, and Subhash C. Dhingra (Consultants). The report, prepared by the task team, benefited from suggestions and comments by members of the project's Advisory Group, composed of World Bank staff, including Patricia H. de Baquero (OPCPR), Chandrasekar Govindarajalu (MNSSD), Peter Johansen (ECSSD), Todd M. Johnson (LCSEG), Jeremy Levin (SASDI), Monali Ranade (ENVCF), and Ashok Sarkar (ETWEN).

This book is based on several country case studies commissioned under the study. The authors of these cases include Brian Henderson (Canada and the United States), Anke Meyer (Germany), Dilip R. Limaye (France and India), Suiko Yoshijima (Japan), Alan D. Poole (Brazil), Xiaoyu Shi (China), and SEVEn (Czech Republic). The team would also like to acknowledge the many energy efficiency practitioners and experts from around the world who shared their valuable time and provided information that contributed to the development of these case studies. The team also acknowledges Pierre Langlois (Econoler), Peter Hobson (EBRD), Russell Sturm (IFC), and Phil Coleman (Lawrence Berkeley National Laboratory) for their global perspectives and insights in support of the study.

Special thanks to Heather Austin (ESMAP) for support in coordinating and facilitating the production and dissemination of the study report. Finally, the team members would like to express their deep gratitude to Amarquaye Armar (ESMAP Program Manager) for all of his strategic guidance and support throughout the study.

Abbreviations and Acronyms

ADB	Asian Development Bank
BEA	Berlin Energy Agency (Germany)
BEE	Bureau of Energy Efficiency (India)
BEEF	Bulgarian Energy Efficiency Fund
CDM	Clean Development Mechanism
CEEF	Commercializing Energy Efficiency Finance
CER	certified emission reduction
CO_2	carbon dioxide
CO_2e	carbon dioxide equivalent
CPWD	Central Public Works Department (India)
DENA	Deutsche Energie-Agentur (Germany)
DH	district heating
DSM	demand-side management
EBRD	European Bank for Reconstruction and Development
ECA	Eastern Europe and Central Asia
ECCJ	Energy Conservation Center, Japan
ECO	Energy Conservation Commercialization Project
EE	energy efficiency
EEEP	Energy Efficiency Enhancement Project
EMC	energy management company

EOC	energy operations contracting
EOI	expression of interest
EPC	energy performance contracting
ESC	energy supply contracting
ESP	energy service provider
ESPC	energy savings performance contract
ESCO	energy service company
ESMAP	Energy Sector Management Assistance Program
EU	European Union
FBI	Federal Buildings Initiative (Canada)
FEMP	Federal Energy Management Program (United States)
GDP	gross domestic product
GEF	Global Environment Facility
GHG	greenhouse gas
GTZ	Gesellschaft für Technische Zusammenarbeit (Germany)
GUDC	Gujarat Urban Development Company (India)
HEECP	Hungary Energy Efficiency Co-financing Program
HVAC	heating, ventilation, and air conditioning
IADB	Inter-American Development Bank
IDIQ	indefinite delivery, indefinite quantity
IEA	International Energy Agency
IFC	International Finance Corporation
IGA	investment grade audit
IPCC	Intergovernmental Panel on Climate Change
IPMVP	International Performance Measurement and Verification Protocol
IQC	indefinite quantity contract
IREDA	Indian Renewable Energy Development Agency
JAESCO	Japan Association of Energy Service Companies
KEMCO	Korea Energy Management Corporation
M&V	measurement and verification
MDB	multilateral development bank
MOE	Ministry of Education
MOF	Ministry of Finance
MTEF	Medium Term Expenditure Framework
Mtoe	million tons of oil equivalent
MWRI	Ministry of Water Resources and Irrigation (Arab Republic of Egypt)
NGO	nongovernmental organization
NPV	net present value

NRC	Natural Resources Canada
NYPA	New York Power Authority
NYSERDA	New York State Energy Research and Development Authority
O&M	operations and maintenance
PICO	public internal performance contracting
PPIAF	Public-Private Infrastructure Advisory Facility (World Bank)
PPP	public-private partnership
REEEP	Renewable Energy and Energy Efficiency Partnership
RFP	request for proposal
TNUDF	Tamil Nadu Urban Development Fund (India)
Toe	tons of oil equivalent
TWh	terawatt hours
UBA	Federal Environmental Agency (Germany)
UESC	utility energy services contract (U.S.–FEMP)
UkrESCO	Ukraine Energy Service Company
ULB	urban local body (India)
UNDP	United Nations Development Programme
UNFCCC	United Nations Framework Convention for Climate Change
USAID	United States Agency for International Development
USDOE	United States Department of Energy
WBG	World Bank Group

Overview

Around the world, energy efficiency is becoming one of the most critical policy tools to help countries meet the substantial growth in energy demand while easing the environmental impacts of that growth. For national governments, energy efficiency is a win-win-win option, providing positive returns to the government, energy consumers, and the environment. Energy efficiency measures are generally viewed as "no regrets" policies, since their net financial cost can be negative—the measures are justified purely based on high financial returns. Unfortunately, despite these promising benefits, achieving significant and sustained efficiency gains has proved daunting in both developed and developing countries.

The public sector holds significant potential for improved energy efficiency globally and represents a large and important market in all countries. The common ownership and homogeneous nature of many of the facilities, particularly those with common functions (schools, hospitals), offer unique opportunities for bundling many projects together, allowing financing at a large scale and attracting new firms into the energy efficiency business. Furthermore, the public sector can have a catalytic effect on local markets by demonstrating good behavior to the private sector and the general public, while stimulating nascent markets for energy efficiency goods and services.

Despite the potential, however, realizing these vast energy savings is hardly trivial. Determining how to package and implement dispersed energy efficiency projects effectively has proved difficult. Although some mechanisms, such as utility demand-side management (DSM) and energy service companies (ESCOs), have been developed for this purpose, experience has shown that the institutional mechanisms must be carefully designed and adapted to fit local needs and situations. The institutional issues are particularly important in the public sector. Overcoming restrictive public regulations, poor incentive structures, limited expertise and information, and other obstacles requires concerted efforts. Although simple measures and universally applicable policies are lacking, experience from a number of countries shows that large-scale energy efficiency gains in the public sector are possible. Governments should pursue a multipronged approach to encourage efficiency improvements in public facilities in all sectors (see Table 1).

Table 1   Typical Barriers and Solutions for Public Sector Energy Efficiency

Barrier	Indicative action and countries
Lack of awareness and information, including costs, benefits, risks, products, and services	Energy efficiency (EE) awareness campaigns, case studies, procurement guidelines, product catalogues/specifications, information dissemination, demonstrations (Brazil, Canada, France, Germany, Japan, Mexico, Sweden, U.S.)
Lack of technical capacity for audits, project design, procurement, implementation and supervision of Energy Savings Performance Contract (ESPC) projects; trust of EE potential	Creation of nodal agencies to provide technical support for EE projects, appointment of energy managers, development of training/educational programs for facility operators and energy managers/Energy Service Providers (ESPs), EE analytical tools, ESPC audit and procurement/contracting guidelines, prequalification of ESPs, Measurement and Verification (M&V) protocols (Brazil, Canada, China, Germany, India, Japan, Mexico, U.S.)
Limited incentives to implement EE (potential loss of budget), try new approaches, take on risks	Revised budgeting to allow retention of energy savings, awards for agencies/staff, EE in management performance reviews, risk sharing/financing programs, EE targets (Brazil, Canada, China, France, India, Mexico, U.S.)
Lack of agency accountability for energy savings	Creation of public sector/agency targets with reporting and monitoring, penalties for nonperformance, energy performance labeling of buildings (China, Germany, Japan, Mexico, Sweden, U.K., U.S.)

(continued)

Table 1   Typical Barriers and Solutions for Public Sector Energy Efficiency *(continued)*

Barrier	Indicative action and countries
Restrictive procurement, contracting, and financing rules	Revised public policies for EE products (e.g., labeled only, life cycle costing) and ESPCs, develop local ESPC models, create public EE funds (Brazil, Canada, China, France, Germany, U.K., U.S.)
Lack of funding for upfront energy audits and project funding	Earmarked public EE budgets, dedicated grant/ subsidy programs, public revolving funds, Demand-side Management (DSM) surcharge or "wire charge," free energy audits by public entities (Brazil, China, Germany, Japan, Mexico, Sweden, Thailand, U.K., U.S.)
Small size and high transaction costs of EE projects	Bundling of public EE projects, model documents and templates to streamline projects, prequalification of ESPs, bulk procurement of EE products (Austria, Canada, Germany, India, Sweden, U.S.)

Source: Authors.

Energy savings performance contracts (ESPCs) have been introduced in many countries to help address some of the more difficult issues associated with facilitating energy efficiency investments. ESPCs have a number of inherent advantages for addressing the specific difficulties that public agencies face. Outsourcing an energy efficiency project in its entirety—from development to financing to monitoring—allows agencies to reap the gains without the hassles of completing each step of the project on their own, often with multiple procurements taking months, if not years. The ability of ESPCs to offer off-budget financing and to pay for themselves from the savings they achieve makes the mechanism even more attractive to public agencies that have small discretionary budgets, or none at all, and a very low tolerance for risk.

Unfortunately, although ESPCs may be well suited to address many of the challenges to improving public sector energy efficiency, rigid public procurement and budgeting guidelines and procedures are quite poorly suited to making ESPC procurement simple, particularly if full project costs and technical parameters have yet to be determined. Furthermore, the complex nature of ESPCs requires significant capacity building throughout the public sector to ensure their successful use.

The study that forms the basis for this book was designed to identify the various approaches and common options used for ESPC public procurement programs and processes, which are based on an international review of detailed country and project case studies. Developed countries studied included Canada, France, Germany, Japan, and the United States. The review also identified program and project examples (both successes and failures) in developing countries, including the Arab Republic of Egypt, Brazil, Bulgaria, China, Croatia, the Czech Republic, Hungary, India, Poland, and South Africa. During this process, some 60 government officials, technical experts, ESCO representatives, and other practitioners were interviewed.

In this book, a private sector service provider is referred to as an "energy service provider," or ESP, and the resulting contract between the public agency and the ESP as the "ESPC"—the energy savings performance contract. The premise of the book is that the ESPC process, carefully developed and customized for the public sector, could bring local and international equipment suppliers and related businesses into the efficiency service business by encouraging them to offer additional technical services, supplier credits, and some technical and/or project performance guarantees, depending upon what services and risks they are willing to undertake. Therefore, this pool of prospective service providers would be broader than typical energy service companies, or ESCOs.

The case studies revealed a number of different approaches by governments to dealing programmatically with the promotion and procurement of ESPCs in the public sector. Whereas some countries, such as Canada and the United States, have initially addressed these issues at the federal level, others, including Germany, focused on developing more local experience first. Some countries, such as the Czech Republic and India, have been heavily influenced by donor programs, while others, including Brazil and Mexico, have not. The various emerging models for dealing with ESPC procurement that the study identified are summarized in Table 2.

Table 2   Emerging Models for Public ESPC Procurement

Model	Description	Cases
1. Indefinite contracting	Umbrella government agency competitively procures one or more ESPs (typically based on general qualifications) and then allows public agencies to enter into direct contracts with selected ESPs without further competition.	Hungary (Ministry of Education), U.S. Federal Energy Management Program (FEMP)
2. Public ESP	ESP is publicly owned, so there is no requirement for a competitive procurement process.	Ukraine (Rivne City)
2a. Super-ESP	A publicly owned ESP contracts directly with a public entity and then subcontracts with smaller ESPs or contractors on a competitive basis.	Belgium (Fedesco), Philippines (EC2), U.S. New York Power Authority (NYPA)

(continued)

Table 2   Emerging Models for Public ESPC Procurement *(continued)*

Model	Description	Cases
2b. Utility ESP	A public entity contracts directly with its utility for EE services without additional procurement (since they are already an existing supplier).	Croatia (HEP ESCO), U.S. utility energy services contract (UESC)
2c. Utility DSM ESP	A publicly owned ESP uses funds from a DSM surcharge to invest in target public agencies at no cost to the agency (so no procurement occurs, since there is no contract/payment).	Brazil
2d. Internal ESP Public Internal Performance Contracting (PICO)	A unit within a public agency acts as ESP, providing technical and financial services, and receives payments through internal budget transfers.	Germany (Stuttgart)
3. Energy supply contracting (*chauffage*)	Public agency contracts out delivery of an energy service, such as lighting or heating, and selects a service provider based simply on cost per unit of service.	Austria, France, Germany
4. Procurement agent	A quasi-public entity or nongovernmental organization (NGO) helps government agencies, often on a fee-for-service basis, develop requests for proposals (RFPs) and assists them through contract award.	Austria, Czech Republic (SEVEn), Germany, Slovak Repub. (CEVO), U.S. New York State Energy Research and Development Authority (NYSERDA)
5. Project bundling	Umbrella government agency bids out a group of buildings or facilities for a large ESPC.	Austria, Germany, India Central Public Works Department (CPWD) and Tamil Nadu Urban Development Fund (TNUDF), S. Africa (Johannesburg), U.S. (California)
6. Nodal agencies	A dedicated energy efficiency agency is appointed to facilitate procurement (prepare model documents, share experiences, provide training, facilitate financing).	India Bureau of Energy Efficiency (BEE), Japan Energy Conservation Center (ECCJ), Korea Energy Management Corporation (KEMCO), U.S. Department of Energy (USDOE)
7. Ad hoc	No explicit program or mechanism to support public ESPCs but no policy to prevent them; projects are developed one at a time.	Arab Repub. of Egypt, Brazil, China, Mexico, Poland, South Africa

When it comes to the specific steps in the procurement process, a number of key issues must be overcome to successfully implement energy efficiency projects using ESPCs. Based on the study, six main steps in the EPSC procurement process were identified, along with key issues and decision points within each step, as presented in Figure 1. This book describes each of these issues in detail and summarizes different approaches and solutions developed by various countries.

Figure 1   Schematic of Typical ESPC Steps and Issues

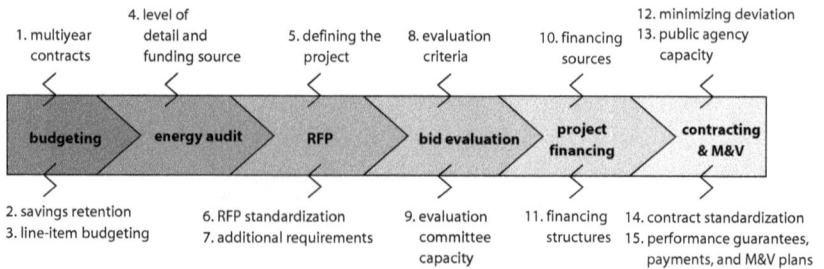

Source: Authors.

Assisting public agencies through the process of ESPC procurement is complex. The multidisciplinary nature of the various issues posed in each step of the process, from budget regulations to energy auditing to public contracting to project financing, makes navigating the process very challenging. And, unfortunately, many of the procurement and budgeting practices vary from country to country, so solutions must be country specific. Nevertheless, as illustrated throughout this book, countries have developed various solutions to deal with all of the issues. Some general recommendations for dealing with some of the main steps in the procurement process are summarized in Table 3.

Table 3   Recommendations for Main Public ESPC Procurement Steps

Main Steps	Recommendations
Budgeting	Consider starting public ESPC procurement schemes with the more autonomous public entities first.
	Gain support from, and work with, parent budgeting agencies.
	After implementing a few ESPCs, to develop public financing programs to help address budgeting, incentive, and financing issues.
	In the longer term, to perhaps require changes to the budgeting laws and regulations.

(continued)

Table 3   Recommendations for Main Public ESPC Procurement Steps *(continued)*

Main Steps	Recommendations
Energy audit	Consider the level of technical information that prospective bidders require to properly define the project. Where appropriate, consider providing basic technical data (facility description, equipment inventory, energy bills, etc.) in lieu of an energy audit.
Bidding documents	Define the project carefully to ensure that it meets local procurement rules and regulations. Consider broader parameters, such as minimum energy savings or target systems, and try to avoid being too prescriptive. Avoid standardizing the procurement documents too early, as that can limit flexibility and the natural evolution of the market. Once a critical mass of projects has been implemented, standardization can facilitate scale-up and reduce transaction costs. Consider adding steps to the bidding process, such as prequalification, detailed audits, pre-bid conferences, and oral presentations, based on local needs and capabilities.
Evaluation process	Consider adopting a two-stage evaluation process in which technical proposals are scored first and the highest-ranked proposals proceed to the financial evaluation stage. Use net present value or another single, comprehensive indicator in the financial evaluation to allow for simple, transparent assessments, while also limiting "cream skimming."
Financing	In mature capital markets, efforts should be made to attract commercial financing for ESPCs with informational and other technical support. Where perceived risks are high, credit or risk guarantee instruments can facilitate accessing commercial financing for ESPCs. In immature markets, particularly where liquidity is an issue, a dedicated energy efficiency fund or entity may be a more suitable approach to support ESPCs. All financing programs should be flexible to allow for maximum market development.
Contracting	Because of the inherent complexities, consider designating certain entities (e.g., nodal agencies, agents, public ESPs) to facilitate public ESPC projects. Once initial ESPCs have performed successfully, develop standardized contracts to further facilitate public energy savings projects. The nature of performance guarantee should be defined based on the type of measure being implemented, and a measurement and verification plan should be included in the contract. The ESPC should address operations and maintenance and client training to ensure that savings persist.

Source: Authors.

For each of the issues in the ESPC procurement process, this book identifies several approaches and, in many cases, presents a continuum of options for countries to consider. By looking at the range of options, a country may be able to find an appropriate and feasible solution based on what has been done elsewhere. Alternatively, a local government or public agency could mix and match, combine, or develop new solutions based on the many approaches presented. As shown in Figure 2, countries may be able to design suitable procurement processes by calibrating the options for each issue. Of course, in reality, the process is never quite so simple. Many of the steps are inextricably linked to one another, making the process more complex. For example, if agency or local procurement regulations require a detailed project definition, then a more detailed initial audit may be required, but the evaluation process may become more straightforward. These interlinking steps also mean that a solution to one issue may limit the possible solutions to another.

Before one becomes too involved in the details of the procurement process, it may be appropriate to first determine possible business and contractual models for ESPs to operate in a particular country or region. Often it may be advisable to begin with simpler models first and develop more complex transactions as the market develops. Familiarity with ESPC models developed in the Western countries can be important in understanding the range of options, but those models need to be adapted incrementally to work in developing countries. Where local ESPC experience exists, it may be prudent to build on successful transactions and institutionalize those aspects that have worked well. It may also be worth considering efforts to bundle projects to reduce transactions costs and make such projects more attractive to larger companies, including international ESPs.

In addition, assessing the relative capacities of the ESP industry and public agency staff is critical to ensure success. The level of sophistication of ESPCs, the ability of local firms to access and provide financing, the willingness of local firms to take on project risks, and other factors must be assessed if a suitable process is to be designed. Efforts should be devoted to developing incremental adjustments to existing procedures and practices, rather than seeking broad changes to local laws and regulations. Some key steps to consider include the following:

1. An *upfront market survey of ESPs* should be conducted to gauge their level of interest in serving the public sector market and their capacity to do so.

Figure 2 Designing a Suitable Procurement Process

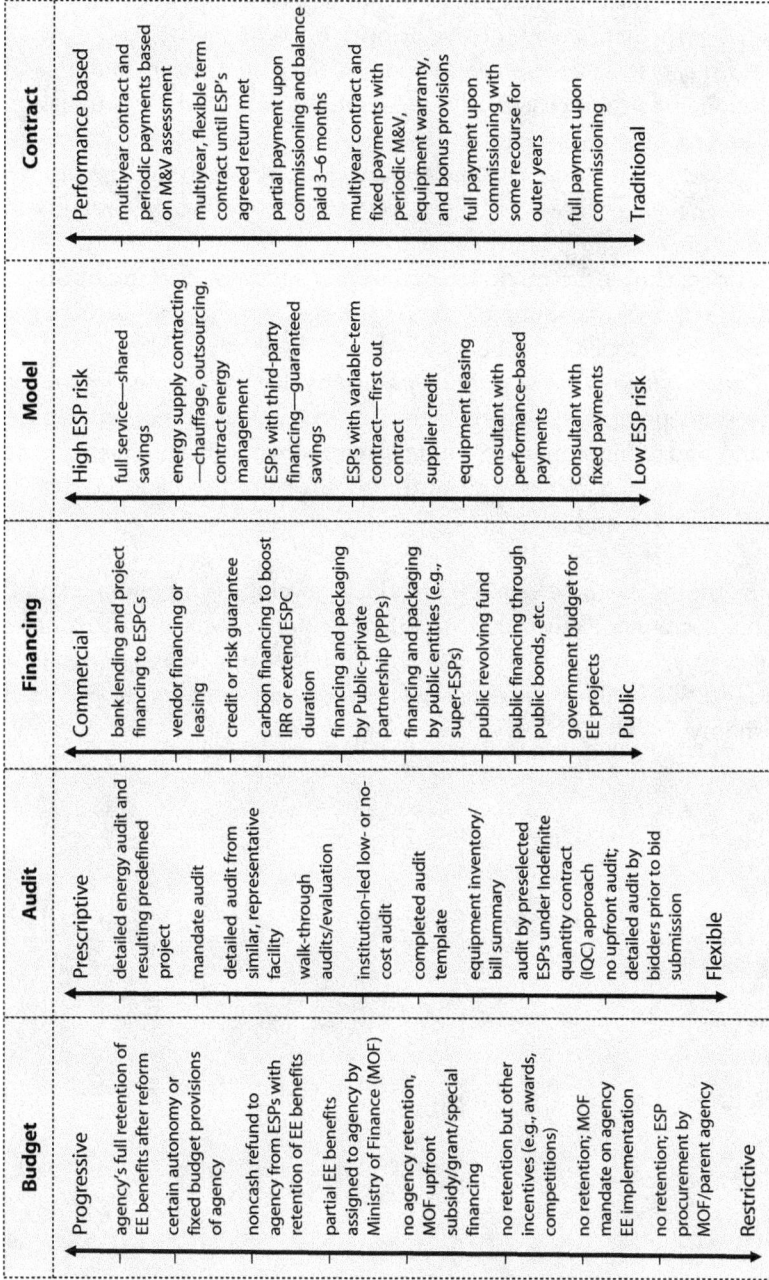

Budget	Audit	Financing	Model	Contract
Progressive	**Prescriptive**	**Commercial**	**High ESP risk**	**Performance based**
agency's full retention of EE benefits after reform	detailed energy audit and resulting predefined project	bank lending and project financing to ESPCs	full service—shared savings	multiyear contract and periodic payments based on M&V assessment
certain autonomy or fixed budget provisions of agency	mandate audit	vendor financing or leasing	energy supply contracting—chauffage, outsourcing, contract energy management	multiyear, flexible term contract until ESP's agreed return met
noncash refund to agency from ESPs with retention of EE benefits	detailed audit from similar, representative facility	credit or risk guarantee	ESPs with third-party financing—guaranteed savings	partial payment upon commissioning and balance paid 3–6 months
partial EE benefits assigned to agency by Ministry of Finance (MOF)	walk-through audits/evaluation	carbon financing to boost IRR or extend ESPC duration	ESPs with variable-term contract—first out contract	multiyear contract and fixed payments with periodic M&V, equipment warranty, and bonus provisions
no agency retention, MOF upfront subsidy/grant/special financing	institution-led low- or no-cost audit	financing and packaging by Public-private partnership (PPPs)	supplier credit	full payment upon commissioning with some recourse for outer years
no retention but other incentives (e.g., awards, competitions)	completed audit template	financing and packaging by public entities (e.g., super-ESPs)	equipment leasing	full payment upon commissioning
no retention; MOF mandate on agency EE implementation	equipment inventory/bill summary	public revolving fund	consultant with performance-based payments	
no retention; ESP procurement by MOF/parent agency	audit by preselected ESPs under Indefinite quantity contract (IQC) approach	public financing through public bonds, etc.	consultant with fixed payments	
	no upfront audit; detailed audit by bidders prior to bid submission	government budget for EE projects		
Restrictive	**Flexible**	**Public**	**Low ESP risk**	**Traditional**

Source: Authors.

2. *Stakeholder consultations* should be held to analyze barriers, assess the types of constraints expected in public procurement of ESPCs, and define the nature and relative priority of the main barriers.
3. A *list of options for surmounting each of the main barriers* should be developed, as an approach to developing possible road maps to navigate the procurement process.
4. It is advisable to *develop and test small procurements*, documenting all the challenges faced, and work collectively to refine the procurement process.
5. Efforts should be made to *expand and replicate*, seeking options to scale up by broadening the range of target systems and increasing the number of facilities to be bundled.
6. *Systems should then be institutionalized* by developing model templates and documents, seeking longer-term changes to public procurement and budgeting systems, creating incentive schemes and financing programs, setting targets and outlining measurement and verification frameworks, and the like.

As the market develops, ESPC models developed and promoted in the public sector are likely to have significant ripple effects in the private sector, as private firms observe what the public sector is doing and participating ESPs begin to market their services and skills to the rest of the economy.

PART I

International Experience in Public Procurement of Energy Efficiency Services

CHAPTER 1

Introduction

According to recent estimates by the International Energy Agency (IEA), the world's primary energy needs will grow by 45 percent between 2006 and 2030, requiring some $26 trillion in investment in supply infrastructure. About 87 percent of this growth is expected to occur in developing countries. Unfortunately, fossil fuels are expected to remain the dominant source of primary energy, accounting for about 80 percent of the overall increase to 2030. China and India, which together will account for about half of this increased demand, will continue to rely heavily on coal to fuel their growth (IEA, 2008).

Developing countries have a critical need for help in meeting their growing energy needs to maintain robust economic growth and social development. However, the recent volatility of oil prices and current projections of increased reliance on oil and gas, much of it coming from the Middle East and Russia, have heightened concern about energy security. Furthermore, increasing concerns over climate change will require that low-carbon options be more actively pursued, with IEA's Reference Scenario showing increases in CO_2 emissions from 28 gigatons in 2006 to 41 gigatons in 2030, an increase of 45 percent.

Why Is Energy Efficiency Important?

Around the world, energy efficiency is rapidly becoming one of the most critical policy tools to help meet the substantial growth in energy demand. By all accounts, energy efficiency programs have always represented a win-win-win option by providing positive returns to the government, energy consumers, and the environment. Such programs can conserve natural resources; reduce the environmental pollution and carbon footprint of the energy sector; reduce a country's dependence on fossil fuels, thus improving its energy security; ease infrastructure bottlenecks and the impact of temporary power shortfalls; and improve industrial and commercial competitiveness through reduced operating costs. In terms of project economics, energy efficiency options are "no regrets" policies, since their net financial cost can be negative—the measures are justified purely based on high financial returns.

Despite these promising benefits, achieving significant and sustained efficiency gains has proved daunting in all countries. As noted in the recent World Bank publication *Financing Energy Efficiency: Lessons from Brazil, China, India and Beyond* (Taylor et. al. 2008), the major constraints on financing and implementing increased energy efficiency are institutional in nature. Determining how dispersed energy efficiency projects can be organized, packaged, financed, and implemented in the most effective and efficient manner has proved very difficult. Some mechanisms, such as utility demand-side management (DSM) and energy service companies (ESCOs) have been developed to address these institutional challenges, but experience has shown that programs must be designed very carefully and adapted to fit local needs and situations. The institutional issues are particularly important in the public sector.

Why Is the Public Sector Important?

Public sector[1] energy use varies widely across countries but generally is a major cost for governments and competes for resources with other critical economic and social development programs. Since the public sector often represents the largest single purchaser of energy within a country, with many 24-hour-load facilities, such as hospitals, universities, orphanages, and so forth, bundling many small and dispersed energy efficiency investments can represent a very attractive business opportunity that can help entice existing commercial suppliers to enter the energy efficiency market. Furthermore, public sector facilities of similar type (e.g., office buildings, schools) tend to have relatively homogeneous end-use consumption patterns, which offer great potential for project replicability. Programs to reduce energy use can reduce energy-related bills, thus creating fiscal space to allow governments to expand social services and meet critical infrastructure investment priorities. Such programs can also ease load demands on often-overstressed utilities, reduce dependence on imported fuels, and reduce environmental impacts from energy use. Energy efficiency investments in public facilities can also be an attractive economic stimulus, by creating local jobs to "green" existing infrastructure while upgrading facilities and lowering future operating costs.

In addition, governments can represent an attractive market segment for commercial equipment and service providers, given their large purchasing requirements and generally low credit risk. The public sector typically represents 10 to 20 percent of gross domestic product (GDP), with much of Europe on the higher end and the United States and Japan on the lower end (Harris 2009). The figure for procurement is much less, but the public sector is still often the single largest purchaser in a country. In the European Union (EU), for example, public procurement is in excess of €200 billion, or about 3 percent of total GDP (PROST 2003). Harmonization of energy efficiency procurement policies across government agencies can thus help lead a shift in the market. In the United States a 1993 executive order directed all federal agencies to purchase only "Energy Star"[2]–labeled, energy efficient computers and office equipment. Although federal sales amounted to only 2 to 3 percent of the market, the order resulted in a substantial increase in manufacturers' requests to join in the Energy Star program, with many types of qualifying equipment quickly reaching penetration rates of 90 percent or more (Harris et al. 2005). In this way, the public sector can lead by example, demonstrating good behavior to the private sector and the general public while stimulating nascent markets for energy efficiency goods and services.

Why Have Results Been So Disappointing?

Implementing energy efficiency projects in the public sector, particularly in developing countries, has been challenging. Efforts to realize substantial gains have been severely constrained by a number of key barriers, including the following:

- Inadequate information about energy efficiency potential and benefits, technologies, products, and practices among public sector facility managers

- Inadequate technical expertise in the public sector on how to conduct energy audits, design and carry out energy efficiency projects, identify quality energy efficiency products, and implement best practices
- Lack of discretionary or dedicated budgets to pay for equipment upgrades, despite attractive payback periods, and limited or restrictive access to appropriate financing
- Mixed or divided institutional incentives for energy savings projects, which can arise when one department pays for equipment and another pays monthly energy bills
- Rigid procurement practices that may not allow life cycle costing, bundling of service/equipment/financing, or use of multiyear contracts
- Low or subsidized energy prices and/or inefficient collection practices

Some of these barriers can be addressed by using the performance contracting approach, in which a public agency engages a commercial service provider to design and implement an energy efficiency improvement project that includes guarantees of energy savings or other performance parameters. The service provider can offer a range of services to the public agency, such as an energy audit; project identification and design; equipment procurement, installation, and commissioning; measurement and verification (M&V); training; and operations and maintenance (O&M). In this way, private sector expertise and capital can be deployed while allowing technical risks to be transferred away from public end users, removing equipment procurement processes from rigid government requirements, and offering more flexible financing options than fixed annual budgeting systems may provide. More important, project development can be outsourced to an entity that has the skills and incentives to overcome any short-term barriers and help realize the significant energy efficiency potential on public premises.

In this book, we refer to the private sector service provider as an "energy service provider" (ESP) and to the contract between the public agency and the ESP as the "energy savings performance contract" (ESPC).[3] The premise of the book is that the ESPC process, carefully developed and customized for public sector programs, could bring local and international equipment suppliers and related businesses into the efficiency service business by encouraging them to offer additional technical services, supplier credit, and some technical and/or project performance guarantees, depending on what services they are willing to offer and what risks they are willing to bear. The public sector provides the added benefit of easier bundling of multiple facilities, thereby creating larger projects that can also attract large and international firms, while making financing needs large enough to attract potential financiers and reduce transaction costs. Such a process could create replicable transaction models that could be further scaled up in multilateral development bank (MDB) and host government energy efficiency operations worldwide.

In addition to procurement-related issues, an integral piece of the puzzle that also needs to be further explored is the type of ESP business model to be considered, as that would affect the procurement processes to be adopted. Considering mixed past experiences with promoting Western-style ESP configurations, we find that business models may need to be custom tailored to suit existing local procurement regulations and market conditions.

Objectives

The primary objectives of this book are to summarize international experience in procuring performance-based energy efficiency services using the ESPC approach in the public sector, to help encourage energy efficiency investments in the developing world, and to identify the different business and procurement models that public agencies may use to engage ESPs. Additionally, the book seeks to do the following:

- Outline the opportunities for ESPCs in the public sectors of developing countries.
- Define an overall procurement process for ESPCs in a public agency.
- Document case studies on specific country experiences using the ESPC approach.
- Suggest more innovative, demand-driven ways for public end users to customize equipment and service solicitations (in terms of end uses, financing requirements, technical/performance guarantees, and so forth).
- Identify key procurement issues related to the ESPC approach and describe how they have been addressed in various countries.
- Offer a road map for employing the ESPC approach and selecting ESPs to implement energy efficiency projects in the public sector.

Structure

This book is based on an international review of country experiences using the ESPC approach in the public sector, along with country case studies in greater depth. Developed-country studies included Canada,

France, Germany, Japan, and the United States. The review also identified project and program examples (both successes and failures) in developing countries, including the Arab Republic of Egypt, Brazil, Bulgaria, China, Croatia, the Czech Republic, Hungary, India, Poland, and South Africa. During the review process, some 60 government officials, technical experts, ESCO representatives, and other practitioners were interviewed. Some of the study findings reflect their perspectives. It should be noted that this book is meant to summarize the procurement practices used by various developed and developing countries to show the variety in their approaches. The World Bank does not endorse any of the particular methods or practices described in the book and maintains its own procurement policies for investment operations with client countries.

Chapter 2 discusses energy efficiency in the public sector, including opportunities, barriers, and typical remedies. Chapter 3 defines the key characteristics of ESPCs and ESPs and summarizes why energy performance contracting makes sense in the public sector. Chapter 4 discusses the various ESP public procurement models that the review identified. Chapter 5 provides an overview of a typical ESPC procurement process and emblematic issues. Chapter 6 identifies and analyzes the main issues in each step of the ESPC procurement process and solutions that have been found for them. Chapter 7 provides some general conclusions and recommendations and presents a road map for implementing the ESPC approach in the public sector. Part II of the book contains summaries of six country case studies and Appendices 1 and 2 contain supporting information, including tables of World Bank and other donor public sector energy efficiency programs.

Notes

1. For the purposes of this book, "the public sector" refers to publicly owned institutions subject to public procurement rules and regulations, including federal/municipal buildings, universities/schools, hospitals/clinics, public lighting, water utilities, public transportation stations, community centers, fire stations, libraries, orphanages, and so on.

2. Energy Star is a U.S. government endorsement label for energy efficient products and practices. For more information, visit http://www.energystar.gov.

3. The private sector organizations that implement projects using the performance contracting approach have traditionally been referred to as "energy service companies" or ESCOs. However, the ESPC approach can be implemented by organizations such as energy suppliers, equipment manufacturers, vendors,

construction management companies, and other related businesses that may not be commonly recognized as ESCOs. The authors therefore refer to the implementers of the ESPC as "energy service providers," or ESPs. The salient characteristics of the ESPC and ESPs are defined in chapter 3.

References

Harris, Jeffrey (Alliance to Save Energy). 2009. "Energy Efficient Public Procurement: Lessons from Around the World." Presentation at European Energy Efficient Public Procurement Workshop, Ispara, Italy, March.

Harris, Jeffrey, Bernard Aebischer, Joan Glickman, Gérard Magnin, Alan Meier, and Jan Viegand, 2005. "Public Sector Leadership: Transforming the Market for Efficient Products and Services." In *Proceedings, ECEEE 2005 Summer Study.* Paris: ADEME Editions. http://www.pepsonline.org/publications/Public%20 Sector%20Leadership.pdf.

IEA (International Energy Agency). 2008. *World Energy Outlook 2008.* Paris: International Energy Agency.

PROST Project (EU SAVE Programme). 2003. *Harnessing the Power of the Public Purse: Final Report from the European PROST Study on Energy Efficiency in the Public Sector.* http://www.eceee.org/EEES/public_sector.

Taylor, Robert P., Chandrasekar Govindarajalu, Jeremy Levin, Anke S. Meyer, and William A. Ward. 2008. *Financing Energy Efficiency: Lessons from Brazil, China, India, and Beyond.* Washington, DC: World Bank.

CHAPTER 2

Energy Efficiency in the Public Sector

Energy Use in the Public Sector

The public sector consumes a fair share of energy and electricity globally. Precise data are often lacking, since many power/heating utilities do not categorize their public sector customers separately from their commercial ones, and what constitutes "public sector" may vary from country to country. According to the data available, however, public sector consumption appears to be in the range of 2 to 5 percent of total energy consumption, perhaps as much as two times more in countries with extensive district heating systems. Chinese government agencies, for example, consume about 5 percent of the country's total energy use and in 2000 spent more than US$10 billion on energy (CECP 2003). In the United States, federal agencies alone consumed 38 million tons of oil equivalent (Mtoe) (1.5 percent) of primary energy in 2006, at a cost of US$17.7 billion, and generated 42.8 million tons of carbon dioxide equivalent (CO_2e) to operate government buildings, vehicles, and equipments (FEMP 2008). If one looks at electricity and heating use only, the share of the public sector is much higher. Almost 9 percent (33 terawatt-hours or TWh) of Brazil's total electricity consumption in 2006 was in the public sector (Meyer and Johnson 2008). About 10 percent of the European Union-15's electricity and heating demand is from the public sector, at a value of about €47 billion in

2001, and 20 percent of electricity and heating loads was attributed to the public sector in the European countries of Estonia, Hungary, Poland, and the Slovak Republic (PROST 2003).

Summary of Energy Efficiency Opportunities by Sector

Although no credible estimates exist of energy efficiency potential in the public sector globally, anecdotal evidence suggests that it is substantial. Public facilities are generally prone to have old and outdated equipment; many pay low energy prices or are not consistently required to pay utility bills. Many public facilities have limited funds to acquire more efficient equipment, which may be costly. Moreover, many facility managers are not aware of energy efficiency products or their performance, and the equipment they have is poorly maintained. In middle-income countries, in facilities where lighting is the primary energy use, energy efficiency savings can be as high as 50 percent. Many public office buildings in developing countries can easily achieve 20 to 40 percent energy savings through retrofits of existing equipment. Savings may not be as significant in many lower-income countries, where access to and use of electricity may be relatively low. The following section summarizes major public end user types and typical energy efficiency opportunities.

Government buildings. Buildings, public and private, consume some 40 percent of global energy and have significant potential for energy savings. Government buildings, particularly in developing countries, tend to be older and to have inefficient and poorly maintained equipment, so their potential for energy efficiency gains is large. Improvements can be made in terms of building envelope measures (windows, insulation), electrical appliances (lighting, heating/cooling, pumping), and office equipment (computers, copiers, printers).[1] New research and pilots to develop energy efficient building designs are currently under way. These approaches seek to maximize natural light ("daylighting"), employ passive solar designs, and incorporate reflective or "green" roofs to reduce lighting and heating/cooling loads. Of course, long-term policies, such as building codes and equipment standards, will yield larger gains over time.

Water utilities. It is estimated that 2 to 3 percent of the world's energy consumption is devoted to pumping and treating water, with potential for energy savings of more than 25 percent (Alliance to Save Energy 2006). Systems in developing countries often have outdated equipment, poor system design, leaks, and other nonmetered water losses. These problems come from a number of underlying factors, such as a lack of investment capital, poor maintenance practices, and limited incentives for energy efficiency. Efficiency improvements can come through reduction of water leakage and waste, downsizing (and rightsizing) pumps, system redesign, pressure management, pump impeller reduction, low-friction pipes, efficient pumps with variable speed drives, load management, power factor improvements, improved maintenance procedures, and so forth.[2] Wastewater treatment plants also provide opportunities for efficiency gains through waste heat recovery and methane capture for power generation, improved pumping, and other measures.

Public lighting. Public lighting is often seen as an essential public service, in terms of both economic activity and improvement in quality of life (e.g., reduction in crime and vehicular accidents). Public agencies commonly use initial cost as the sole deciding factor in the procurement decision and thus do not look to recurring and future costs. Therefore, many developing countries have a preponderance of mercury vapor lamps, which can typically have capacities of 100 to 400 watts. Efficiency improvements may be gained by replacing these inefficient lamps with more energy efficient ones (e.g., metal halide, T-5 fluorescent tube lamps, or high-pressure sodium vapor lamps).[3] Improved lamps can also decrease light trespass (extraneous light on adjacent properties), reduce the number of street lighting poles required, and improve light color quality. Street lamp retrofits can save up to 30 to 40 percent of energy costs, last three to five times longer, and have payback periods of less than three years. In addition to one's switching out the lamps, time clocks and automatic control systems, combined with redesign of street lighting systems to eliminate over-lit and under-lit areas, may help achieve further energy and cost savings.[4]

Institutional facilities. Institutional facilities, which can include colleges and universities, schools, hospitals and clinics, libraries, museums, orphanages, and other such organizations, also have great energy reduction potential. For such facilities, use of multifunctional buildings, innovative waste recycling and waste-to-energy schemes, cogeneration (or combined heat and power), water reuse, and load management can dramatically improve energy efficiency, particularly in facilities with 24-hour loads. As in the case of government buildings, efficiency improvements can be significant in lighting, electrical appliances, medical or laboratory equipment, and building envelope measures.

Barriers to Energy Efficiency in the Public Sector

Energy efficiency has long been a challenge in the public sector, even in developed countries. Despite the opportunities identified above, implementation of energy efficiency measures is constrained by rigid public sector procurement practices that focus on first costs and a lack of discretionary budgets to make investments in energy efficient equipment. There is also a principal-agent or split-incentive issue, in that a parent budgeting agency may determine capital budget needs, while the subordinate agency is responsible for paying the monthly energy bills. Other constraints include the following:

- Government agencies are not typically as responsive to price signals since they lack a commercial orientation.
- Public procedures for equipment and service procurement are not flexible.
- Constrained annual budgets make funding for capital upgrades difficult, while restrictions on public financing and typical one-year budget appropriations make it difficult to amortize costs.

An expanded summary of typical barriers by stakeholder is provided in Table 2.1.

Energy Efficiency Programs in the Public Sector

Country governments have undertaken a wide variety of interventions to address barriers to energy efficiency improvements in the public sector, including the following:

Policy measures:
- *Energy pricing* (time-of-use pricing, feed-in tariffs for co-generation, etc.)
- Energy efficient *product procurement* (e.g., public sector minimum energy performance standards, purchase of labeled products only, life cycle cost evaluation, bundling agency purchases to lower costs)
- Setting and monitoring of *energy efficiency targets* in public facilities, with mandatory submission of annual plans
- Allowing use of *energy savings performance contracts* (ESPCs)
- *Building codes* and certification

Table 2.1 Typical Barriers to Energy Efficiency in the Public Sector

Policy / Regulatory	Public end user	Equipment/ Service provider	Financiers
• Low energy pricing and collections • Rigid procurement and budgeting policies • Limitations on public financing • Ad hoc planning	• Limited incentives to save energy or try new approaches • No discretionary budget for upgrades or special projects • Unclear about ownership of energy/cost savings • Limited availability of project financing • Lack of awareness and technical expertise • Behavioral biases	• Higher transaction costs for public sector projects • Perceived risk of late/nonpayment by public sector • High project development costs • Limited technical, business and risk management skills • Limited access to equity and project financing	• High perceived public credit risk • New technologies and contractual mechanisms • Small project sizes/high transaction costs • Behavioral biases

Source: Authors.

Procedural changes:
- *Changes in budgeting* to allow facilities or agencies to retain energy savings
- Designation of an agency or facility *energy manager*
- Periodic *energy audits* to identify cost-effective energy efficiency measures
- *Operations and maintenance changes,* such as automatic shutoff of lighting and other equipment during evening and weekend hours

Informational programs:
- *Standard bidding documents* and templates
- Dissemination of *analytical tools,* such as energy and life cycle cost calculators
- Establishment of *benchmarks*
- *Guidelines and good practices* for building management, including energy management
- Public sector energy efficiency *case studies* and *newsletters*
- *Training* of public sector staff, facility managers, and procurement officers

Incentives:
- Funding for *energy audits*
- *Public financing* for energy efficiency retrofits and upgrades, often at low interest rates
- *Awards* for high-performing public facility managers, agencies, and cities
- Publishing *agency performance achievements,* ranking and rating agencies

Most countries' experiences suggest that a broad set of policies helps to send signals to the market that energy efficiency is a priority and that the government is taking actions to support it. However, some recent experiences suggest that too many policies and regulations can actually confuse agencies and the market and encounter diminishing returns. While many of the mechanisms that focus on information and awareness are helpful, oftentimes the policies, incentives, financing, and procedural changes are equally critical, if not more so, in helping to achieve meaningful investments and results within the public sector.

World Bank Group Public Sector Energy Efficiency Portfolio

The World Bank has been investing in energy efficiency in the public sector for decades. Its efforts have typically included financing of public infrastructure upgrades and expansion seeking to incorporate more energy efficient and modern designs than would have been included otherwise. World Bank lending has also focused heavily on supply-side energy efficiency, largely in state- and municipal-owned electric power and heating utilities, to reduce technical losses and improve quality of service and financial performance, as well as on creating better commercial incentives through sector reforms.

More recently, the World Bank and its private financing agency, the International Finance Corporation (IFC), have developed dedicated energy efficiency financing programs, often with significant Global Environment Facility (GEF) cofinancing, to support local lending markets. Although generally targeted to industrial and commercial sectors, a number of the programs have also financed public sector energy efficiency improvements where a commercial service provider, such as an energy service provider, was involved. Such projects were often municipal-level investments, such as street lighting and government buildings. The World Bank has also begun efforts to strengthen its pipeline for energy efficiency investments using carbon financing, where verified reductions in CO_2 emissions can be converted into certified emission reductions (CERs) with a corresponding revenue stream.

Appendix 1 of the book, provides a snapshot of World Bank Group projects approved for fiscal years 2000 to 2009 that include demand-side energy efficiency investments in the public sector. Many of these components are parts of much larger investments or financial intermediation programs. It is, therefore, not always possible to separate out the energy efficiency component to assess the aggregate Bank financing in this area. The list is meant to be representative only, as it does not seek to include every World Bank–financed project in public sector construction that could be considered energy efficient. Nor does it include supply-side projects, which represent core utility lending and are thus already mainstreamed within the World Bank's business development.

As this representative project list shows, the World Bank has approved 22 projects that explicitly include public sector energy efficiency components. Of these investments, 8 focused on public (office) buildings, 7 on schools and/or hospitals, 5 on municipal water supply systems, 3 on housing, and 3 on street lighting. It is interesting to note that the vast majority of the projects (17, or 77 percent) are located in the World Bank's Eastern Europe and Central Asia (ECA) Region. More than half of these ECA operations deal with modernization of heating systems and building weatherization. Although reducing costs is a motivating factor in these operations, modernization of the energy systems and improved service quality and comfort levels are often more prominent drivers. There are also five ECA projects that explicitly address energy efficiency in water utilities, although many more projects that deal with water utility modernization and expansion entail implicit efficiency improvements that are not routinely specified in appraisal documents. It is also observed that these projects are split fairly evenly between urban and energy sector operations, pointing to the cross-sectoral nature of energy efficiency in the public domain.

During the same period, the IFC mobilized GEF funds to support three energy efficiency financing/technical assistance programs with public sector components, of which two were multicountry programs. These projects included components for street lighting, schools, and/or municipal buildings. Most of them were in Eastern Europe. The World Bank has also expanded its carbon finance operations to public sector energy efficiency in its client countries. Two energy efficiency projects in India, one dealing with water supply and the other with street lighting, are in the advanced stages of project preparation. Carbon revenues from these energy efficiency investments are expected to help improve the benefit-cost profile of the projects, while reducing project risk with this additional source of revenues.

Within the World Bank Group's portfolio, only about six projects explicitly promoted the ESCO mechanism to package and implement energy efficiency projects, and only a handful were outside the ECA Region. Therefore, one can conclude that public energy efficiency opportunities, particularly outside this region, remain substantially untapped.

Other Public Sector Energy Efficiency Programs

In addition to World Bank investment programs, several other donors, as well as many developing country governments, have launched energy efficiency programs targeting the public sector. These have included large-scale investment programs by regional development banks, such as the European Bank for Reconstruction and Development (EBRD; see Box 2.1) and the Asian Development Bank (ADB), as well as more technical assistance–oriented efforts by United Nations agencies and bilateral donors such as the U.S. Agency for International Development (USAID) and the German Gesellschaft für Technische Zusammenarbeit (GTZ) in collaboration with German energy agencies. Within the past several years, other organizations have also become involved with energy efficiency in the public and municipal sectors, including the William J. Clinton

Box 2.1

The European Bank for Reconstruction and Development (EBRD) and the Public Sector

EBRD has perhaps more experience working on energy efficiency in the public sector than any other donor, the result of very high level management commitment early on to support energy efficiency investments, as well as the demands of the region in which it operates.

Like many donors, EBRD had some early setbacks. In the mid-1990s, there was a widely held misconception that international ESCOs would be eager to enter developing country markets and, in doing so, take on all project risks and provide technical know-how, financing, and so on. Unfortunately, many early tenders (such as EBRD's 500-building bid in Lodz, Poland) failed to attract a critical mass of qualified and attractive proposals. Without access to local and affordable financing, reasonable prospects of clients paying on time, strong enforceable contracts, standardized procedures for verifying savings and resolving disputes, and the like, international ESCOs were not eager to bid on these early projects, and locally established firms had neither the expertise nor the capital base to take on project financing or offer performance guarantees.

Since then, EBRD has substantially refined its approach to dealing with these issues. It developed and implemented a very successful project in Bulgaria, which allowed a local ESCO, Enemona, to sell its receivables to a local bank backed by EBRD financing, constructing projects with short-term capital and then accessing the longer-term capital for refinancing. The bank has replicated this model in Russia. In 1998, EBRD helped to create UkrESCO, a joint stock ESCO in the Ukraine to implement turnkey energy savings projects. An initial loan of US$20 million to UkrESCO was fully disbursed, and a second tranche was approved in 2005.

EBRD has concluded that the good elements of ESPCs, such as multiyear contracts, use of savings to repay the investment, and ability to retain the savings, represent common problems but require locally tailored solutions for the programs to succeed. Equally important is the need to develop simple contracting solutions that allow local companies, many of which may not be well capitalized or be able to offer full performance guarantees, to participate in such projects. EBRD is now working in a new GEF-supported project in Russia to bring its vast previous experience to the development of a locally tailored program for public buildings.

Sources: EBRD 2007, 2008; Peter Hobson, EBRD, pers. comm.

Foundation (Clinton Climate Initiative) and the Renewable Energy and Energy Efficiency Partnership (REEEP). A representative (but by no means exhaustive) list of these programs appears in Appendix 2.

Several observations can be made about this representative sample of 24 programs launched from 2000 to 2009:

- About two-thirds of the programs are in the ECA region.
- EBRD, USAID, and the United Nations Development Programme (UNDP) have been more active donors in the area.
- About a third of the projects have involved the creation of some financing program or fund.
- More than half had ESCOs as an explicit instrument to support project development and implementation.
- Fourteen of the projects focused on schools and hospitals, 6 on public buildings, 4 on water systems, 4 on housing, and 2 on street lighting.

According to a number of interviews with the project sponsors, it appears that the programs with the greatest reported impact in energy cost savings have had technical assistance components linked to financing programs.

UNDP has been active in this area under its GEF programs, with public sector energy efficiency projects in Bulgaria, Hungary, Russia, the Slovak Republic, Ukraine, and Vietnam. All of the projects are geared toward technical assistance, although some of the later ones have included loan guarantees or funds (e.g., Russia, Slovak Republic). All the projects in the ECA region include some component to support ESCO development, and the project in Ukraine includes the creation of a public ESCO in Rivne City (UNDP 2002) to provide financing for upgrades of municipal heating customers. As a result of the decentralized structure of UNDP, each program is country specific, and there appears to be no common institutional approach at present.

The Asian Development Bank has only recently become involved in energy efficiency in the public sector, outside of the more typical supply-side investments that are its conventional business. In early 2009, the ADB board of directors approved the Philippines Energy Efficiency Project which, among other things, would facilitate energy efficiency retrofits in public buildings (ADB 2009). The project includes a component to establish a super-ESCO, called "EC2," under the Philippine National Oil Company, a government-owned corporation, to develop, finance, and implement projects in the public sector and support ESCO development through aggregation in the private sector. By acting as an ESCO in the public sector, the ADB hopes to alleviate many typical procurement barriers. To avoid creating a monopoly project, the ADB envisions subcontracting work to smaller, private sector ESCOs, mostly on a fee-for-service basis.

Notes

1. For example, see Levine et al. (2007), chapter 6 of the Intergovernmental Panel on Climate Change (IPCC) report, for more information on buildings, http://www.ipcc.ch/pdf/assessment-report/ar4/wg3/ar4-wg3-chapter6.pdf.

2. See the Alliance to Save Energy's 2007 *Watergy Handbook* for more discussion on opportunities for water and energy efficiency in water utilities, at http://www.watergy.net/resources/publications/watergy.pdf.

3. Light emitting diodes (LED) are now being used for municipal traffic lights and are being tested for street lighting. Although the initial cost of LEDs is still high and technology still pre-commercial, the savings of more than 50 percent and substantially longer life can make them more economically viable in the near future.

4. See NYSERDA's "How-to Guide to Effective Energy-Efficient Street Lighting: For Planners and Engineers," October 2002, at http://www.rpi.edu/dept/lrc/nystreet/how-to-planners.pdf.

References

ADB (Asian Development Bank). 2009. "Philippine Energy Efficiency Project. ADB Report and Recommendation of the President to the Board of Directors." ADB Project No. 42001. January.

Alliance to Save Energy. 2006. "Watergy Case Study: Emfuleni Municipality, South Africa." http://www.watergy.net/resources/casestudies/emfuleni_southafrica.pdf.

CECP (China Certification Center for Energy Conservation Products). 2003. "Brief Report on Energy Consumption Audit in Chinese Government Agencies." http://www.pepsonline.org/publications/CECP%20Executive%20Summary.pdf.

EBRD (European Bank for Reconstruction and Development). 2007. "Bulgaria ESCO Fund." EBRD Investment Report, EBRD 2007.

———. 2008. "Improving Efficiency in Public Buildings in the Russian Federation under the Energy Efficiency Umbrella Program." GEF Project Document, 2008.

FEMP (U.S. Federal Energy Management Program). 2008. *FY2006 Annual Report to Congress on Federal Government Energy Management and Conservation Programs.* http://www1.eere.energy.gov/femp/about/annual_report.html.

Levine, M., D. Ürge-Vorsatz, K. Blok, L. Geng, D. Harvey, S. Lang, G. Levermore, A. Mongameli Mehlwana, S. Mirasgedis, A. Novikova, J. Rilling, and H. Yoshino. 2007. "Residential and Commercial Buildings." In *Climate Change 2007: Mitigation.* Contribution of Working Group III to the *Fourth Assessment Report of the Intergovernmental Panel on Climate Change*, ed. B. Metz, O. R. Davidson, P. R. Bosch, R. Dave, and L. A. Meyer. Cambridge, UK: Cambridge University Press.

Meyer, Anke S., and Todd M. Johnson. "Energy Efficiency in the Public Sector—A Summary of International Experience with Public Buildings and Its Relevance for Brazil." ESMAP Report, World Bank, Washington, DC, May 2008.

PROST Project (EU SAVE Programme). 2003. *Harnessing the Power of the Public Purse: Final Report from the European PROST Study on Energy Efficiency in the Public Sector.* http://www.eceee.org/EEES/public_sector.

UNDP (United Nations Development Programme). 2002. "Ukraine—Climate Change Mitigation in Ukraine through Energy Efficiency in Municipal District Heating (Pilot Project in Rivne) Stage 1." Final GEF project document.

CHAPTER 3

Opportunities for Energy Savings Performance Contracts in the Public Sector

What Is an Energy Savings Performance Contract?

An energy savings performance contract (ESPC) involves providing an energy consumer, or "host facility," a range of services related to the adoption of energy efficient products, technologies, and equipment. The services provided may also include the financing of the energy efficiency upgrades, so that the host facility has to put up little or no capital. The host facility pays for the services from the money it saves from reduced energy consumption. In many cases, the compensation is contingent on demonstrated performance, in terms of energy efficiency improvement or some other measure, thereby creating a system where the services and equipment can be paid from the actual energy cost savings.

The ESPC is implemented by service providers that have traditionally been referred to as "energy service companies," or ESCOs. ESPCs and ESCOs were first developed in the United States in the late 1970s in the wake of the energy crisis and the rapid increase in oil prices resulting from the OPEC oil embargo and the Iranian revolution.[1] Increasing energy prices created awareness among building and industry owners and managers of the need to use energy more efficiently. The elements of performance contracting evolved as the ESCO industry developed, and these concepts have now been accepted as standard features of ESCO services. Traditionally the term "ESCO" has been used to designate an organization that provides a full range of energy services.[2] However, the ESPC approach can be implemented by organizations such as energy suppliers, equipment manufacturers, vendors, construction management companies, engineering firms, mechanical and electrical contractors, and other related businesses that may not be commonly recognized as ESCOs. Furthermore, in developing countries, where full-service ESCO models may be too complex or impractical, simpler models may be preferable to Western-style ESCOs. Therefore, in this book we use the term "energy service provider" (ESP) to designate any organization that may offer any of a range of services associated with energy efficiency project implementation using the ESPC approach.

Key Characteristics of an ESPC

The ESPC approach is generally characterized by the following key attributes:[3]

- ESPCs can offer a complete energy efficiency service, including design, engineering, construction, commissioning, and operations and maintenance (O&M) of the energy efficiency measures; training; and measurement and verification (M&V) of the resulting energy and cost savings.
- ESPC services also include providing or arranging financing, often with a link between ESP compensation and project performance, so that customers pay for the energy services with a portion of actual energy cost savings achieved.
- ESPCs typically include performance guarantees, based on the level of energy or energy cost savings, for the entire project (as opposed to individual equipment guarantees offered by equipment manufacturers or suppliers).
- Most of the technical, financial, construction, and performance risks are borne by the energy service provider under the ESPC.

ESPC Models

Although specific approaches to the ESPC vary, they can generally be characterized into two basic types of agreements[4]—"*shared savings*" and "*guaranteed savings.*" In both models, the ESP provides the complete range of implementation services and generates energy and cost savings. The differences are in the ways that the project is financed, payments are made from the host facility to the ESP, and energy and cost savings are allocated between the ESP and the host facility.

In the *shared savings* model, the ESP generally provides, or arranges for, most or all of the financing needed to implement the project. The ESPC specifies the sharing of the cost savings between the ESP and the host facility over a period of time. The ESPC may last 3 to 10 years (but may be as short as one year in some underdeveloped markets and 15-plus years in developed ones), with the sharing of payments structured such that the ESP will recover its costs and obtain the desired return on its investment within that period. The host facility generally makes no investment in the project but receives a share of the savings during the contract period and all of the savings after the contract period. It thus maintains a positive cash flow throughout the life of the project. A shared savings agreement must include a prespecified protocol for M&V of the actual savings achieved.

In a *guaranteed savings* agreement, the host facility generally takes the needed loan on its own balance sheet. The ESP guarantees achievement of certain performance parameters (such as efficiency, energy savings, or cost savings) in the ESPC, which also specifies the methods for measurement and verification. Payment is made once the project performance parameters have been confirmed. The guaranteed savings agreement generally provides the ESP a fixed payment or payment stream upon the satisfaction of the performance guarantee, and it may also provide the ESP an incentive payment if actual performance exceeds the guaranteed level. On the other hand, if the savings fall below the pre-agreed levels, the ESP would be obligated to cover the host facility's loan repayments until project performance had been restored. Figure 3.1 illustrates the two models.

An earlier and somewhat different model of the ESPC that has been in use in France for more than 60 years is the *chauffage* model. Also referred to as "contract energy management," "energy outsourcing," or "energy supply contracting," this model is one in which the ESP takes over O&M of a customer's energy-using equipment and sells the output (e.g., steam, heating and cooling, lighting) to the customer at an agreed price (Bertoldi and Rezessy 2005). This model is a form of outsourcing in which the costs for all equipment upgrades, repairs, and so forth are borne by the ESP, but ownership typically remains with the customer. The fee that the client pays under a chauffage arrangement is calculated based on its existing energy bill, minus a percentage savings (often in the range of 3 to 10 percent), or a fee may be charged per square meter of conditioned space. Under the chauffage arrangement, the client is guaranteed an improved level of energy service and a reduced energy bill. Contracts of this type tend to be substantially longer than others, ranging from 10 to 30 years.

It should be noted, though, that many ESPs using chauffage contracts focus primarily on energy production at a lower cost and improved O&M

Figure 3.1 ESPC Business Models—Shared and Guaranteed Savings

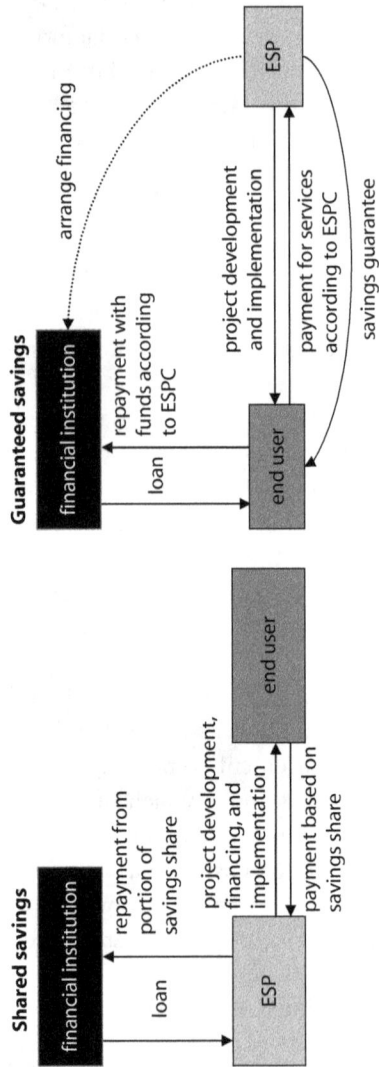

Source: Taylor et al. 2008.

rather than on investments in energy efficiency. It is also important to point out that in France, the chauffage model has not been applicable for energy efficiency investments in the public sector because French law precludes investment by the private sector in public facilities. But a new law on public-private partnerships enacted in 2004 greatly facilitates the use of ESPCs in the public sector.

In much of the developed world, these three contracting models remain the most prevalent ones. Since the ESPC model has been introduced in developing countries and adapted to fit local circumstances, however, the variation of models has grown (see Box 3.1). Often the focus has been on making the model simpler, since many developing countries cannot handle the complex and lengthy contracts and sophisticated M&V methodologies. In some cases, new approaches have emerged and evolved, and it is fully expected that this evolution will continue in the future.

Again, for the purposes of this book, an ESPC can include most of these models or variations thereof. Participating ESPs must create a package

---

Box 3.1

Examples of Different Energy Service Company (ESCO) Business Models

The list ranges from the full service/high-risk contracts to low service/risk.

Full-Service ESCO. The ESCO designs, finances, and implements the project, verifies energy savings and shares an agreed percentage of the actual energy savings over a fixed period with the customer. This is also referred to as the "shared savings" approach in the United States.

End-Use Outsourcing. The ESCO takes over operations and maintenance of the equipment and sells the output (e.g., steam, heating/cooling, lighting) to the customer at an agreed price. Costs for all equipment upgrades, repairs, and so on, are borne by the ESCO, but ownership typically remains with the customer. This model is also sometimes referred to as "chauffage," "contract energy management," or "energy supply contracting."

ESCO with Third-Party Financing. The ESCO designs and implements the project but does not finance it, although it may arrange for or facilitate financing. The ESCO guarantees that the energy savings will be sufficient to cover debt service payments. This is also referred to as "guaranteed savings" in the United States.

(continued)

Box 3.1 *(continued)*

ESCO Variable Term Contract. This is similar to the full-service ESCO, except that the contract term can vary based on actual savings. If actual savings are less than expected, the contract can be extended to allow the ESCO to recover its agreed payment. A variation is the "first-out" model, in which the ESCO takes all the energy savings benefits until it has received its agreed payment.

Equipment Supplier Credit. The equipment supplier designs and commissions the project, verifying that the performance/energy savings matches expectations. Payment can either be made on a lump-sum basis after commissioning or over time (typically from the estimated energy savings). Ownership of the equipment is transferred to the customer immediately.

Equipment Leasing. Similar to supplier credit, the supplier receives fixed payments from the estimated energy savings. However, in this case the supplier owns the equipment until all the lease payments, and any transfer payments, are completed.

Technical Consultant (with Performance-Based Payments). The ESCO conducts an audit and assists with project implementation. The ESCO and customer agree on a performance-based fee, which can include penalties for lower energy savings and bonuses for higher savings.

Technical Consultant (with Fixed Payments): The ESCO conducts an audit, designs the project, and either assists the customer with implementing the project or simply advises the customer for a fixed, lump-sum fee.

Source: Adapted from World Bank 2005b.

from the array of possible services, as previously described. The exact package of services would depend on the public agency's needs and the local market's capabilities, but some of the basic ESPC elements should be maintained for the public agency to reap the full benefits; that is, (a) ESPCs should tie at least a portion of the ESP remuneration to project performance, and (b) ESPCs must involve project implementation and not just upstream aspects, for example, only energy audits or equipment sales, or only downstream services, such as maintenance.

International Experience with ESPCs and ESPs

A number of recent reports have reviewed and described the characteristics, activities, and accomplishments of ESPs (most of the reports refer to them as ESCOs) in different countries (MOTIVA 2005). These reports point out the extensive interest throughout the world in the concept of ESPCs and the many policy initiatives and programs that national governments, multilateral development banks, and other donors have undertaken to encourage and promote them. A key conclusion of the reports is the need to further promote the ESPC approach to facilitate the implementation of energy efficiency projects.

What makes public sector procurement different is that any firm, whether or not it considers itself an energy service company, can participate in the process as long as it provides the services requested. In this way, the pool of potential bidders can be greatly expanded to include a very wide range of ESPs. However, as is discussed later in the book, the ESPC must be designed based on the capabilities of the prospective ESP bidders, the types of services they are able and willing to provide, and risks they are able and willing to assume.

Why ESPCs Make Sense in the Public Sector

As noted in chapter 2, energy efficiency investments in the public sector in developing countries have been limited by several barriers. ESPCs can help address some of those, as outlined in Table 3.1.

It is not suggested here that ESPCs are a "magic bullet" to solve the deep and complex issues related to improving energy efficiency in the public sector. However, many of the core elements of ESPCs are quite attractive as means to deal with prevailing obstacles to achieving energy efficiency gains in the public sector:

- ESPs can bundle all of the various steps typically required to implement an energy efficiency project into one contract, thereby reducing transaction costs and hassles for the public agency, while easing some of the more restrictive public procurement requirements.
- ESPs can transfer technical risks away from public end users and financiers by providing performance guarantees that assure customers of successful implementation of energy efficiency measures and offer O&M services to ensure that the installed equipment continues to perform at high efficiency levels.
- ESPs can provide broad and deep technical and logistical capabilities to implement projects, as well as provide training and capacity building to the public agency staff.
- ESPs can facilitate access to external capital and offer more flexible financing options for projects, thereby alleviating some of the budgetary constraints that public agencies typically face.

Table 3.1   How ESPCs Can Address Key Barriers

Key public sector barriers	How ESPCs can help
Lack of commercial incentives to reduce operating costs	ESPCs cannot deal with fundamental lack of incentives but can help reduce transaction costs and perceived risks by offering a full package of services and assuming project performance risk.
No incentive to save energy (no retention of savings)	ESPCs cannot address the principal-agent issue but can better define the costs and benefits upfront, so agencies can negotiate and apportion them appropriately.
High perceived risks from new technologies and mechanisms	ESPCs include performance guarantees to assign many project risks away from the public agency and financier to the ESP.
Inflexible procurement procedures	ESPCs can allow for high returns by offering the best value projects to be implemented, bypassing procurement of service providers and equipment for each measure.
Constrained annual budgets for capital upgrades	ESPCs often offer project financing, either through an ESP or a third party, with repayment derived from project savings.
Small projects with high project development and transaction costs	ESPCs allow for multiple smaller public projects to be bundled, since there is common ownership, often with notional audit/baseline information, thus helping to address development and transaction costs.
Inadequate information and technical know-how	ESPCs invite technically competent private sector firms to compete based on their qualifications, experience, and best project ideas.

Source: Authors.

For these reasons, it is worth studying how such mechanisms can be more readily adopted in developing countries. To find that out, the global review conducted for this project posed nine questions to government officials and experts about their experiences with energy savings performance contracts:

1. Is an audit performed before the ESP is selected? If so, who pays, who conducts it, and what level of detail is required? If not, what technical information is provided in the request for proposal (RFP), and who compiles it?
2. Can the public facility allow for different types of projects (e.g., lighting, HVAC, motors and pumps, control systems) to be bid? Can they accept multiple offers?
3. What evaluation criteria are used to select the ESP?
4. Does the RFP require a multiyear contract? If so, how is this point dealt with, given the typical one-year public budget cycles? What are typical procurement lead times, and how can they be shortened?

5. Are ESPs required to provide financing? If so, what are the sources and terms? If not, how is the project financed?
6. What type of performance guarantee, if any, is required? What insurance, guarantees, or recourse is provided if the project does not perform?
7. What M&V procedures are required? How are these defined and enforced? Is there a third-party verifier if disputes arise, and if so, what type of organization provides this service?
8. How are incentives provided to public sector agencies and staff, by policy or other instruments, to undertake energy savings performance contracting if any energy savings may result in lower future appropriations? Can specific staff be rewarded for such programs?
9. Are there local institutions that (could) take on the function of helping to package public sector projects for bidding, bundle them, perform contracting and financing negotiations, and assist with commissioning and monitoring? If so, what has inhibited wider scale replication? What has been the experience—what lessons have been learned, if any—with bundling of public projects?

Using the answers to these questions and the case studies, we identified several programmatic models, as described in the next chapter.

Notes

1. See also World Bank 1999.
2. The services include preliminary feasibility analysis, detailed audits of facilities, identification of energy efficiency options, engineering, procurement, construction management, installation services, performance guarantees, financing, O&M, training, and performance M&V.
3. Adapted from World Bank 2005a.
4. Many reports on international ESCO activities have described these two basic models. See for example, MOTIVA (2005); and Bertoldi and Rezessy (2005).

References

Bertoldi, Paolo, and Sylvia Rezessy. 2005. "Energy Service Companies in Europe: Status Report," European Commission, Joint Research Centre, EUR 21646 EN.

MOTIVA. 2005. *International Review of ESCO Activities.* http://www.esprojects
.net/en/energyefficiency/financing/esco/publications/.

World Bank. 1999. *"The Energy Service Industry: The Experience of the United States and Canada,"* Energy, Mining and Telecommunications Department Occasional Paper No. 12. Washington, DC: World Bank.

World Bank. 2005a. "World Bank Public-Private Infrastructure Advisory Facility, A Strategic Framework For Implementation of Energy Efficiency Projects for Indian Water Utilities."

World Bank. 2005b. Energy Efficiency Portfolio Review and Practitioners' Handbook, Washington, DC: World Bank.

Taylor, Robert P., Chandrasekar Govindarajalu, Jeremy Levin, Anke S. Meyer, and William A. Ward. 2008. *Financing Energy Efficiency: Lessons from Brazil, China, India, and Beyond.* Washington, DC: World Bank.

CHAPTER 4

Main Public Energy Savings Performance Contract Procurement Models

Emerging Models for Public ESPC Procurement

The case studies revealed a number of different approaches to dealing with the complex set of issues surrounding public ESPC procurement. Some countries, such as Canada and the United States, sought to address them at the federal level early on, whereas others, such as Germany, focused on developing more local experience first. Some countries, such as the Czech Republic and India, have been heavily influenced by donor programs, while others, including Brazil and Mexico, have not. It is also interesting to note that the countries with the most developed ESPC markets, including Canada, Germany, and the United States, have all relied extensively on the public sector market to stimulate the development of the Energy Service Provider (ESP)/Energy Service Company (ESCO) industries.

Perhaps more interesting than these approaches is that this review identified some emerging models for dealing with the issues associated with public procurement of ESPCs. In some cases, the model is quite country specific, while in others the model or variations can be found in many countries. A summary of these models, including countries where examples occur and their advantages and drawbacks, appears in Table 4.1. Each of the models is discussed in more detail next.

Table 4.1 Emerging Models for Public ESPC Procurement

No.	Model	Description	Cases	Pros	Cons
1.	Indefinite contracting	Umbrella government agency competitively procures one or more ESPs (typically based on general qualifications) and then allows public agencies to enter into direct contracts with selected ESPs without further competition.	Hungary Ministry of Education, U.S. (USDOE/FEMP)	This approach allows much easier contracting between smaller public agencies and ESPs, with lower risks for upfront audits and project design costs.	Upfront contracting can create barrier to market entry during contract period, nontransparent direct contracting approaches, and less leverage for public entities to negotiate on price.
2.	Public ESP	ESP is publicly owned, so there is no requirement for a competitive procurement process.	Ukraine (Rivne City)	This model can reduce transaction costs for procurement and provide greater access to concessional international financing (through sovereign loans), while raising comfort level of public agencies that know little about EPCs.	Public ESPs can become monopolistic and may not provide services as efficiently and cost-effectively as private ESPs; this approach may not lead to a sustainable, vibrant ESP market; needs viable exit strategy.
2a.	Super-ESP	A variation of the public ESP model, a publicly owned ESP, contracts directly with a public entity and then subcontracts with smaller ESPs or contractors on a competitive basis.	Belgium (Fedesco), Philippines (EC2), U.S. (NYPA)	This model has the advantages of a public ESP while still allowing more private ESP participation and competition.	This approach can create an artificial barrier between the contractor and customer; has very limited track record in developing countries; allocation of risk to contractors is tricky.
2b.	Utility ESP	A public entity contracts directly with its utility without additional procurement (since it is already a existing supplier).	Croatia (HEP ESCO), U.S. (UESC),	This model allows bundling of energy and efficiency services with existing utility, public or private, and easier repayment through utility billing; utilities can access cheaper financing.	It can create monopolistic utility ESP, there may be some conflict between efficiency and energy supply services.
2c.	Utility DSM ESP	A publicly owned ESP uses funds from a DSM surcharge to invest in target public agencies at no cost to the agency (so no procurement since there is no contract/payment).	Brazil	This approach can address the procurement and financing barriers together.	Using DSM surcharges for public agencies can create perception of unfairness; ESP may not be service oriented without contractual relationship; often there is no performance guarantee.

44

2d.	Internal ESP (PICO)	A unit within a public agency acts as ESP, providing technical and financial services, and receives payments through internal budget transfers.	Germany (Stuttgart)	There is no procurement and the transaction is internal to the public administration.	As with other public options, efficiency relative to commercial ESP is not clear.
3.	Energy supply contracting (chauffage)	Public agency contracts out delivery of an energy service, such as lighting or heating, and selects a service provider based simply on cost per unit of service.	Austria, France, Germany	This model is well demonstrated and has a very simple procurement and contracting approach(no performance contract).	It often requires long terms (20–30 years) to be viable; focus on supply-side gains only; better suited for central systems (heating, cooling).
4.	Procurement agent	A quasi-public entity or NGO helps government agencies, often on a fee-for-service basis, develop RFPs and assists them through contract award.	Austria, Czech Republic (SEVEn), Germany, U.S. (NYSERDA), Slovak Repub. (CEVO)	This approach is more market based and allows agent to evolve approaches as the market develops.	It can lead to monopolistic behavior as agent has no incentive to share approaches; agent may not have ability to change public policies; developing countries may not have logical agent candidates.
5.	Project bundling	Umbrella government agency bids out a group of buildings or facilities for a large ESPC.	Austria, Germany, India (CPWD, TNUDF), S. Africa (Johannesburg), U.S. (California)	This approach favors competition while expanding project sizes and thus reducing transaction costs.	Public agencies have less control in how their projects are bundled; large project sizes may inhibit market entry of new (or local) ESPs.
6.	Nodal agencies	A dedicated energy efficiency agency is appointed to facilitate procurement (prepare model documents, share experiences, training, facilitate financing).	India (BEE), Japan (ECCJ), Repub. of Korea (KEMCO), U.S. (USDOE)	This approach provides a strategic review of procurement programs and sharing of experiences and model documents.	Nodal agencies may have limited ability to influence procurement, budgeting policies; assistance is general and often excludes detailed transaction support.
7.	Ad hoc	No explicit program or mechanism exists to support public ESPCs but no policy to prevent them; therefore, projects are developed one at a time.	Arab Repub. of Egypt, Brazil, China, Mexico, Poland, South Africa	It allows for full innovation and development of demonstrable approaches before developing guidelines and model documents.	The transaction costs for the first project(s) are very high and some reinventing of the wheel in early projects is likely.

Source: Authors.

Note: DSM = demand-side management, PICO = public internal performance contracting, NGO = nongovernmental organisation, RFP = Request for Proposal.

Indefinite Quantity Contracts

Indefinite quantity contracts (IQCs) are contracts in which the general goods and services required are defined but the quantity is not. The Federal Energy Management Program (FEMP) in the United States uses such contracts frequently when it is anticipated that multiple subagencies will be procuring similar types of services but do not know exactly when and to what extent.[1] To avoid each subagency's procuring the same services each time, a parent agency can issue a blanket award to one or more energy service providers on a competitive basis, without specifying the quantity of services to be delivered.[2]

ESPs are selected based on their demonstrated capabilities to manage the development and implementation of multiple ESPCs over large geographic areas. The selected providers can then propose a specific project to a subagency and be awarded a contract on a sole-source basis to implement it. (Sole source selection under the USDOE/FEMP indefinite quantity contracts was the typical mode of ESCO selection from 1998 to 2008, but this is no longer the case with the new, 2009 IQC awards.) The World Bank also has a form of IQCs, referred to as "indefinite delivery contracts," when borrowers need to have specialized services on call but cannot define the extent and timing in advance.

Indefinite quantity contacts can be very efficient, because they can allow the parent agency to bundle many smaller assignments upfront and negotiate better terms with the ESPs. They can also greatly reduce transaction costs by avoiding many smaller procurement packages, which can be particularly difficult for smaller agencies (see Box 4.1 for an example from Hungary). Also, since there is less competition for the subawards, it is easier for selected ESPs to negotiate energy audits and detailed project proposals directly, without excessive concerns over proprietary information, initial audit costs, and the like. But IQCs can create barriers to market entry for new ESPs by locking up all contracts to the selected firms for a fairly long period (often five to seven years in the United States); that may not be a major issue in developed countries but could be in developing ones. Concerns have also been raised over the transparency of making subawards to one of the selected firms with little or no competition. Larger agencies have less leverage to negotiate costs if cost structures have already been set under the blanket IQC contract.

IQCs typically have two parts. Under such schemes, the ESP first conducts a detailed audit and then, if the audit is accepted by the public client, implements the project and is paid for both the audit and project from the savings. If, however, the public client decides not to proceed,

Box 4.1

Indefinite Quantity Contracts (IQC) in Schools in Hungary

In 2006, the Hungarian Ministry of Education (MOE) launched a large-scale education facilities renovation finance program (called the Szemunk Fenye program) to modernize lighting and heating systems in schools. The program used an innovative approach for a pooled public tender process to competitively select a single consortium of firms using an ESPC. The umbrella tender also required bidders to propose unit prices for heating and lighting systems.

The winner of the tender was a consortium led by the OTP Bank and included a Hungarian ESCO, Caminus. MOE signed a 20-year framework agreement with the consortium which, under an umbrella IQC-type contract, would provide energy efficiency improvement services to municipalities and other school operators without an additional tender process for each individual project. Schools are not obligated to use the selected consortium, but if they do not, they must follow the usual competitive bidding process.

The goal of the MOE is to complete lighting and heating system modernization projects of at least HUF 11.1 billion (US$50 million) per year between 2006 and 2010. The total expected size of the project is HUF 56 billion (US$250 million). The winning consortium received a partial credit guarantee from the International Finance Corporation for up to US$250 million to help it access local lending. As of August 2008, the total value of contacts signed was HUF 5.0 billion (about US$22 million).

Source: IFC, 2006; Tibor Kludovacz and Russell Sturm, IFC, pers. comm.

then it is obligated to pay the ESP for the full cost of the audit. In this way, there is no risk to the ESP of not being paid for the audit.

Some models also use the "open book" contract method, in which the ESP simply negotiates with the client the costs for its services, often on an hourly or daily basis, and a reasonable markup for equipment and other subcontracts. In this way, the ESP acts more as an agent than a typical ESCO, since it generally does not take on all of the project performance risks. The advantage is that the public agency does not have to worry about an ESP charging exorbitant rates after being short-listed under an IQC. The ESP can be required to shop around to secure the best equipment, financing, and other pricing for the public client. However, such an approach has drawbacks, in that it may not include a full performance guarantee; such models also are not suited to ESPs that are also equipment suppliers. This approach has been used in Canada, Croatia, and the United States, sometimes under the rubric of a public ESP (see next section).

Public ESPs

Many governments have taken a more active role in promoting ESPCs, sometimes by actually creating publicly owned, or partially public-owned, ESPs. Often this was done to promote ESPs (or ESCOs) in general, examples being China (pilot EMCs in Beijing, Shandong, and Liaoning—see Box 4.2), Ukraine (UkrESCO), and Poland (MPEC). These were typically formed when the local ESP markets were nascent and some public effort was deemed necessary to catalyze them. Although

---

Box 4.2

Energy Performance Contracting in China

In 1996, with support from the World Bank and Global Environment Facility (GEF), three pilot ESCOs (or energy management companies, "EMCs," as they are referred to in China) were established in the Beijing municipality and Shandong and Liaoning provinces. The companies were essentially created by the government, with company shares owned by state and provincially owned enterprises (e.g., energy conservation and technical service centers, industries, utilities) but operated commercially for profit. All three EMCs operated as "full-service" ESCOs; that is, they identified, designed, financed, and implemented projects, with their payment based at least in part on actual energy savings. By 2006, their combined annual ESPC investments reached about US$30 million. More important, during this same period, the EMC industry grew to more than 400 companies, with a combined investment of more than US$1 billion in 2007 alone.

The three pilot EMCs largely targeted the industrial sector, which represented much of the energy savings potential, but they also completed projects in the commercial and public sectors. Their unique status and business model allowed them to enter into many direct contracts with public entities, typically based on the submission of a technical proposal but without formal competition. About 10 percent of their 300 projects between 1998 and 2005 were in public entities, including utilities, universities, hospitals, and municipal buildings.

Source: Shi 2008.

some are only partially owned by the government, in most cases they are required to operate as fully commercial companies. In a few more recent cases, countries have worked to create ESPs specifically to address some of the issues associated with ESPs in the public sector. An example is the ESCO in Rivne City, Ukraine, which was established with United Nations Development Programme (UNDP)/GEF support and subsequently opened up to private equity partners. Often in these latter cases, no competitive process is required, since a public agency is simply contracting with another public entity.

Such an approach can help reduce transaction costs associated with complex service procurement, allow for multilateral development bank financing of the ESPs (since they are publicly owned), and help concentrate ESPC expertise in a few agencies. However, a danger exists that these ESPs can exhibit monopolistic behavior and may not provide services as efficiently as fully private ESPs. If the objective is to stimulate a competitive ESP market, such public ESPs may need to have a strategy to phase out public ownership or foster development of new ESPs in the market.

Some variations on this approach include the following:

Super-ESP. With a super-ESP, a publicly owned and operated ESP contracts with a public agency (usually without competition) for an energy efficiency project and then subcontracts with smaller private ESPs or contractors on a competitive basis to actually complete the work. This approach can help to stimulate a private ESP market but can create an artificial separation between the contractors and the client. Further, allocation of project performance risk from the super-ESP to the subcontractors can be difficult. There are many examples of super-ESPs, including the New York Power Authority (NYPA) in the United States (see Case Study 1), Fedesco in Belgium (see Box 4.3), and the EC2 in the Philippines.

Utility ESP. In this model, an ESP is created from a publicly owned (or partially public) power utility or district heating utility to provide services to its customer base. The Croatia HEP ESCO (see Box 4.4) and the U.S. Utility Energy Services Contract (UESC) are good examples of this approach. An advantage is that the ESP can be selected or appointed without a separate procurement, since the utility already has an implicit contract with the public agency to provide energy services. Such an approach can also benefit from enhanced credibility (when the utility has a good reputation), access to customer energy use

Box 4.3

The Belgian Super-ESP—Fedesco

Fedesco was established in 2005 by the Belgian Federal Investment Company as a public limited company to finance energy and water efficiency measures in 1,800 federal government buildings. The overall objectives of Fedesco are to reduce greenhouse gas emissions and federal energy bills. Fedesco sponsors and funds energy audits to identify potential energy efficiency measures and provides the financing needed for implementation. This includes investments in the building envelope and technical installations, as well as other services, and behavior campaigns. Fedesco collaborates extensively with the Federal Building Agency, which owns and manages most government buildings. Fedesco's investments in energy efficiency improvements are repaid from the energy cost savings on a "first-out" basis. In 2008, the Belgian federal government established a goal for Fedesco to reduce CO_2 emissions from federal buildings energy use by 22 percent by 2014. The investment approved to meet this goal is €210 million over five years. It is estimated that 45 percent of the investment will be through ESPCs.

Sources: Paolo Bertoldi, Benigna Boza-Kiss, Silvia Rezessy, 2007; http://www.fedesco.com.

data, and the ability to bundle smaller projects with the lower-cost financing that utilities can often access. It can also allow repayments though utility billing systems, where allowed. As with the other models under this subset, however, it can engender monopolistic behaviors and can sometimes create conflict between the utility's energy saving and energy supply business models.

Utility Demand-Side Management (DSM) ESP. This approach is similar to the utility ESP scheme, except that the utility uses its own funds, typically from a DSM surcharge (sometimes also known as a "public benefits charge"), to cover the energy efficiency improvements in a public agency. Brazil provides an example of this model (Poole 2008). Since the utility covers all the costs, there is no contract and thus no need for any procurement. Obviously, this ESP can address upfront procurement and financing barriers, but it is unlikely to be sustainable and is difficult to implement transparently and fairly (as demand will undoubtedly exceed supply). With no contract, the utility may not have an incentive to be client oriented.

Box 4.4

A Utility ESP in Croatia

Croatia has faced a growing imbalance of energy demand and domestic supply over the past decade. Very inefficient use of energy, due in part to years of underinvestment and decapitalization, aggravated this problem. It was estimated that energy efficiency improvements could save the country 25 percent of its energy use and save the government more than US$50 million in energy costs each year.

In 2003, the World Bank and GEF approved a project to create an ESCO subsidiary within the national power utility, HEP. HEP ESCO was capitalized by a World Bank loan, HEP equity, local banks, and other sources to offer energy efficiency services to public and private clients. GEF funds were also mobilized to provide additional credit enhancement for HEP ESCO projects and provide some technical assistance to the ESCO and local banks. Since the Croatian market was so small, the government did not foresee a risk of crowding out the private sector, and HEP ESCO uses the "open book" model to keep its pricing fair and transparent. Government entities can directly contract with government companies and their subsidiaries, so public agencies are not required to conduct any competitive procurement to contract with HEP ESCO. To date, about 186 million kuna (US$35.4 million) in energy savings contracts have been signed.

Source: "Croatia Energy Efficiency Project Appraisal Document," 2003; Peter Johansen and John Cowan, World Bank, pers. comm.

Internal ESP. Another form of public ESP is the public internal performance contracting, or PICO, model. Under this scheme, a unit within a public administrative agency, such as the technical department of a municipality, delivers technical and financial services, and the remuneration takes place through cross-payments of budgets (under a shared savings arrangement) from the client department's reduced energy costs to the PICO unit (within the same public administration) (PROST 2003). The city of Stuttgart, Germany, successfully completed more than 130 PICO projects from 1995 to 2000 representing €2.6 million. Such a model can greatly facilitate ESPC projects, inasmuch as there is no procurement involved and both parties are internal. As with other public options, however, there may be drawbacks resulting from the incentives for a public service provider to be as effective and efficient as a commercial one.

Energy Supply Contracting

Energy supply contracting, also referred to as "contract energy management," "chauffage," or energy outsourcing, is a well-developed mechanism (for example, in Austria, France, and Germany) in which a public agency contracts out the delivery of an energy service, such as heating, and selects an ESP largely based on the cost per unit of service. This selection can be much simpler than evaluating and contracting typical ESPCs, where evaluation is more complex and the contract involves a performance guarantee. Typically, energy supply contracts require longer terms (10–30 years) to be viable and are best suited for centralized systems, such as heating and cooling. Such contracts generally focus only on reducing supply costs, whether on-site or off-site, and they often ignore demand-side efficiency gains since the interest of the ESP is to sell more energy services and not energy efficiency per se.

Procurement Agent

In several countries, procurement agents have been developed to facilitate ESPCs in the public sector. In Germany (see Box 4.5), the government had substantial involvement in their creation (e.g., BEA, DENA); in the Czech Republic, the government was not involved (e.g., SEVEn). Agents can be public agencies, public-private partnerships (PPPs), nongovernmental organizations (NGOs), or private consulting firms; they assist public agencies, typically on a fee-for-service basis, through the entire procurement process (from the energy audit through contract award and measurement and verification, or M&V). This approach can be viewed as less government invasive. It allows agents to develop their own solutions in a more market friendly manner. However, agents may not always have strong incentives to share their experiences and lessons learned. Such agents may also have limited ability to influence changes in public policies, where the policies are deficient, and many developing countries may not have obvious candidates with the requisite technical expertise to serve as public ESPC procurement agents.

Project Bundling

Many countries, regardless of the specific approach they use, have found that common ownership allows the bundling of many smaller public projects into a single procurement. Typically, this ownership is done through a parent or umbrella agency, although it has also been done in many

Box 4.5

Berlin Energy Agency and Energy Savings Partnership

The Berliner Energieagentur GmbH (BEA) was founded in 1992 by the State of Berlin, Germany, in a public-private partnership with a federally owned bank and utilities, to promote energy efficiency and renewable energy. BEA provides consulting services (e.g., audits and analyses, project development, tender management, project management and advisory services, training, etc.) to businesses, public authorities, and nonprofit organizations. BEA also acts as an ESCO in the field of small-scale combined heat and power and solar energy.

Under its Energy Savings Partnerships, BEA selects and pools several buildings (from about four to as many as 400), typically with minimum annual energy bills of around €200,000, for ESPCs. BEA works with its clients to bid out energy savings projects, which include energy supply contracts, for ESCOs to design, implement, and finance the projects. In this way, BEA acts as a procurement agent, providing technical advice and intermediary services between the ESCOs and clients in all stages of the project, from the initial diagnosis (energy audit) to project monitoring. BEA has also developed model ESPCs and guidance notes for contracting with ESCOs, in particular, the well-known Hessen ESPC guidelines, which were written by BEA in 1997. By early 2008, over 20 "energy saving partnerships" had been implemented in more than 500 facilities and about 1,300 public buildings, using the ESPC model.

Source: Meyer 2009.

countries with multiple municipalities or even agencies within different departments. Examples include the ESPC procurements conducted by the Central Public Works Department and the Tamil Nadu Urban Development Fund in India, and the City of Johannesburg, South Africa.

Bundling increases the aggregate size of projects, which usually makes them more attractive to ESPs and lowers the transaction costs for each facility. In developing countries, bundling can send a powerful signal to private firms that there are large opportunities in the public sector ESPC business, which may encourage more international and local firms to enter the market. However, the larger transactions can be more complex, and each individual agency will undoubtedly have less control of the overall contract. In developing countries, very large contracts may prevent smaller firms from being able to participate, particularly local firms with limited capital.

Nodal Agencies

Some countries have not explicitly addressed the public procurement issue but have designated nodal agencies to support energy efficiency programs in general. Some of these agencies have begun to address public ESPC barriers directly through a range of programs, such as the development of model bidding documents and contracts, sharing early experiences, training of agencies and ESPs, facilitation of financing, case studies, and so on. Others, such as the Korea Energy Management Corporation (KEMCO), actually provide financing for public sector ESPC projects. Other examples include the U.S. Department of Energy, India's Bureau of Energy Efficiency (BEE), and Japan's Energy Conservation Center (ECCJ). Such an approach supports existing institutional arrangements for energy efficiency projects in general, and the nodal agency can at least take a more strategic view of market development efforts while sharing early experiences among public agencies. Unfortunately, nodal agencies may not be able to sufficiently influence procurement and budget policies, may not have all the skills needed to address the range of issues associated with ESPC public procurement, and typically are not staffed to provide more than just general advice, which may not be enough to see through initial procurements.

Ad Hoc

A number of countries lack explicit programs or policies to promote energy efficiency service procurement in the public sector but have no major policy or other impediments to prevent such projects from being undertaken. Such countries often have a handful of successful transactions in the public sector but no mechanism to replicate them widely. This arrangement may make sense for many developing countries in the early years, as various approaches are being tried and tested and as yet no preferred model or process exists. Early innovation can often be encouraged without disseminating model documents too early in the development of the market. However, transaction costs for each project, in the absence of any government guidelines or sample templates, are likely to be higher. Some "reinventing the wheel" for each project may occur that can slow replication. The Arab Republic of Egypt, China, Mexico, Poland and South Africa provide examples in this category.

Summary of Model Options

These programmatic procurement models, which have emerged from the review of international experience, are but a few of the many possible approaches to promoting public sector ESPCs. However, the range of models is instructive, as it provides examples of the many actions a government can take, depending on the level of market intervention, financial support, and risk it is willing to undertake. For example, should the government wish to promote ESPCs more actively, taking on a more visible role and some market risks, a publicly owned ESP can be a suitable policy option. For those governments that prefer a less intrusive role, focusing on facilitating deals through procurement agents and bundling may be more appropriate. It should be noted that these approaches are not mutually exclusive; a government can use IQCs to promote ESPCs, for example, while also supporting procurement agents and use of project bundling.

In addition to these models, each step of the ESPC procurement process requires further review and analysis, which can be found in the next two chapters.

Notes

1. The U.S. Federal Energy Management Program refers to these as "Indefinite Delivery/Indefinite Quantity" (IDIQ) contracts.
2. In Canada and the United States, where this approach is commonly used by federal agencies, multiple ESPs are selected under the IQC. A subagency may negotiate an ESPC with any one of the selected ESPs under the IQC or may compete the project within the list of the ESPs.

References

C40 Cities Climate Leadership Group. 2009. "Building Efficiency—Berlin." http://www.c40cities.org/bestpractices/buildings/berlin_efficiency.jsp.

Energie Cities. 2003. "Energy Performance Contracting in Pools-Berlin." http:www.energie-cities.org/db/berlin_566_en.pdf.

IFC, "The Role of IFC in Financing Building Renovations in Hungary," presentation at DEEM Seminar, Budapest, September 20–21, 2006.

Meyer 2009; Michael Geissler (BEA), "Energy Efficiency and Energy Services: Experiences of Berliner Energieagentur (BEA)," presentation to World Bank Energy Efficiency Thematic Group, Washington, DC, March 2008.

Paolo Bertoldi, Benigna Boza-Kiss, Silvia Rezessy, *Latest Development of Energy Service Companies across Europe,* European Commission, Joint Research Centre, 2007; also http://www.fedesco.com.

Poole, Alan. 2008. "The Procurement of Energy Efficiency Performance Contracts in Brazil's Public Sector." Unpublished World Bank case study.

PROST Project (EU SAVE Programme). 2003. *Harnessing the Power of the Public Purse: Final Report from the European PROST Study on Energy Efficiency in the Public Sector.* http://www.eceee.org/EEES/public_sector.

World Bank 2003, "Croatia Energy Efficiency Project Appraisal Document," World Bank Report No.25592-HR, September 11, 2003.

Xiaoyu Shi, "Energy Performance Contracting in China: A Case Study," unpublished World Bank case study, 2008.

CHAPTER 5

Public Energy Savings Performance Contract Procurement Steps and Key Issues

Overview of the ESPC Process

The major steps in the procurement process for an energy savings performance contract, or ESPC, are not very different from most public procurement models:

- Conducting an energy audit
- Developing the bidding documents or request for proposal (RFP)
- Evaluating and selecting the energy service provider (ESP)
- Mobilizing financing
- Contracting and measurement and verification (M&V)

This section briefly discusses these major steps and introduces the challenges associated with public ESPC procurement. Although the steps are generally consistent with World Bank procurement guidelines, the purpose of this chapter is to present what various countries have done in their procurement processes, some of which may not be fully consistent with those that the World Bank recommends. Chapter 6 includes a detailed discussion of the steps, issues associated with each of them, and remedies from various countries.

The Energy Audit
The energy audit is an inspection of a building or group of buildings to assess energy consumption, define the baseline, identify methods for reducing energy use, and evaluate the potential for energy cost saving measures. This inspection can include identification of changes to a building's structure, equipment, systems, and operational procedures that will result in less energy consumption and provide sufficient cost savings to pay for the improvements over a given number of years.

This step can essentially serve as a prefeasibility study for an energy efficiency project by confirming that cost-effective energy savings opportunities exist, identifying target systems to be retrofitted, and defining the parameters of the project that will be stated in the RFP. Different types of energy audits go into varying levels of detail, and efforts should be made to tailor the audit to the specific needs of the host facility.

The Request for Proposals
In general, the development of the bidding documents may involve the following steps:

- Prequalifying ESPs
- Defining the project and services to be provided
- Preparing the RFP
- Pre-bidding conference, site visit, and oral presentations

Prequalifying ESPs. The prequalification step may be appropriate to screen interested bidders and ensure that those invited to submit detailed proposals have adequate capabilities and resources to prepare quality bids and successfully carry out the intended work. Prequalification (as opposed to short-listing) requires that the applicants meet a minimum set of specific, objective criteria. All applicants that meet these criteria are then invited to bid. Short-listing, as the name implies, restricts the field of bidders to a fixed number (usually four to six). The World Bank generally recommends prequalification to ensure that unqualified firms are spared the high cost of preparing detailed bids and recommends short-listing only for consultant procurement. Some countries have sought to combine the two by developing a short list of qualified firms. Under typical schemes, a public agency issues an invitation for prequalification or a request for expressions of interest (EOI), the latter for short-listing.[1] The agency then evaluates the qualifications information that firms submit and, using prespecified evaluation criteria (e.g., key personnel, past experience, ability to mobilize financing), develops a prequalified or short list of firms, which then receive invitations to submit detailed proposals.

Defining the project. Defining the project in the RFP is a challenging and critical step in the process. Whereas the purpose of an ESPC is often to allow ESPs to offer their best solutions for the current energy systems within the host facility's premises, some basic parameters need to be established and contained in the RFP. These parameters can include target systems, minimum energy savings, sharing of savings, and so on, as well as services required to be included (e.g., engineering and project design, procurement and installation, financing, measurement and verification, and operations and maintenance).

Preparing the RFP. The request for proposals is the most important bidding document in the procurement process, as it defines the project requirements, evaluation criteria, contractual provisions, and other parameters that will guide the process and eventual contract. The RFP document must be custom designed to fit local conditions and agency preferences and may vary considerably from one country to another, sometimes even from one project to another within the same country. The RFP typically includes instructions to the bidders, specifies the information and format required for proposals, presents basic technical information about the facilities the project will deal with, provides a "scope of work" that describes the work to be done, includes general terms and conditions and a draft contract, and includes other particulars of the agency's requirements. The RFP document can be quite complex as a result of the various technical, financial, and contractual requirements.

Other steps. Other steps that are often included in the RFP process include upstream consultations with potential bidders, a pre-bidding conference (to discuss the contents of the RFP and respond to questions), site visits (to allow bidding ESPs to gather additional performance information on the target facilities), and oral presentations (to allow those with the highest rankings to present their proposals and answer questions from an evaluation committee). These steps are optional and should be based on the needs of the public agency issuing the RFP and the capabilities and experience of the pool of bidders.

Bid Evaluation

Once the public agency has received the proposals, it must assess them based on the criteria specified in the RFP. ESPC projects are very complex because of their combination of technical, financial, project implementation, and performance monitoring requirements, and because the agency must assess multidimensional technical and financial proposals. Technical proposal evaluation may be more straightforward, because it is based on aspects that are contained in most service contracts, such as the methodology, work plan, and staffing. However, assessing the financial proposals can be complicated, inasmuch as there is no single price; proposals will include multiple indicators, all of which are relevant, such as the investment amount, total energy and cost savings, share of savings to be allocated to the agency, duration of the contract, life of the equipment, and so on. In the end, the firm that is selected should provide the best value to the agency.

Financing

A key element of the ESPC process is mobilizing financing for the energy efficiency project. In a typical ESPC arrangement, the RFP requires bidding ESPs to offer a plan for financing the project. Unfortunately, in underdeveloped markets bidding ESPs may have trouble raising financing on their own, and therefore some government-sponsored financing program may be necessary. In other cases, ESP financing may be more expensive than public sources such as municipal debt, and for that reason financing may not be required in the RFP. Financial structuring of a project can be complex, so the agency needs to consider upfront which entity will take on the debt and assume responsibility for repayment.

The Contract

Once the highest-ranked ESP has been identified, it is invited to negotiate the final contract. Whereas this process may be straightforward for many types of contracts, it is more complicated for ESPCs because of the many technical, financial, and legal parameters and the possible lack of experience of the tendering agency. A further complication is that the final ESPC provisions may not be known until an investment grade audit, or IGA, has been completed. Only after the energy service provider has completed the IGA, which establishes the project's baseline, and it is accepted by the public client will the agency then negotiate the specific terms of a final ESPC. The contract must also include a measurement and verification (M&V) plan.

The M&V provisions are a very important part of the ESPC process, since they generally determine the payments made to the ESP. The plan may be specified in the RFP, but it may also be proposed by the bidding ESPs. In either case, the ESP must develop detailed M&V protocols by the completion of the investment grade audit, and these must be agreed with the public agency. This final measurement and verification plan is then incorporated into the final energy savings performance contract. Many agencies and ESPs use the M&V protocols specified in the International Performance Measurement and Verification Protocol (EVO 2007). These protocols may be adjusted appropriately by mutual agreement for the specific energy efficiency measures being installed.

Key Issues Posed by ESPCs in the Public Sector

Despite the promising attributes noted previously, ESPCs have not been widely used in the public sectors of developing countries. The reasons are complex and multifaceted. In many countries, an initial focus on establishing local ESP industries envisioned that these companies could then promulgate the ESPC model in all sectors. Unfortunately, many countries lacked the legal and financial infrastructure to adapt to and support such complex business models. New ESPs generally either lacked the technical and operational expertise to carry out all the functions typically associated with ESPCs or lacked the balance sheets to mobilize the financing that such business models require. Local ESPs often had no track record in the market to carry out sophisticated projects, while international ESPs, with better expertise and access to capital, were usually not keen to invest in these emerging markets because of a host of risks (e.g., perceived small markets and projects, unclear legal and regulatory regimes, concerns about client creditworthiness, lack of access to appropriate local project financing, and the like). Developing countries also have limited equity markets and few investors willing to create new companies and test new business types.

Within the public sector rigid procurement and budgeting guidelines often prevent public institutions from engaging ESPs, particularly where full project costs and technical parameters have yet to be determined (see Box 5.1 for key issues in Brazil). The review identified six main steps to the adoption of ESPCs in developing public sector markets, along with specific issues and decision points within each step (Figure 5.1):

Budget provisions for ESPCs
1. Multiyear contracts
2. Retention of energy savings
3. Line-item budgeting

Box 5.1

Procurement Challenges with ESPCs in Brazil

Brazil's federal-level legal framework governing public sector procurement is found in the general law for procurement (Law 8666 of 1993), as well as in legislation on the budgeting process (Law 4320 of 1964) and Complementary Law 101 of 2000 (known as the Law of Fiscal Responsibility). There are equivalent state-level laws on the same lines. The requirements in these laws that are most critical for ESPC procurement are summarized below:

1. **Project description.** Brazilian law requires RFPs to include a "basic project" (*projeto básico*) that defines the work to be contracted and demonstrates its feasibility—technical, economic, and environmental. In addition, the RFP must demonstrate that the implementation costs are compatible with resources available to the public agency, show that the relevant technical alternatives were evaluated, and confirm that the most favorable one was selected. To comply requires carrying out a detailed audit to define the "basic project," which would be costly and would reduce the possibility for innovative solutions to be proposed by bidding ESPs.

2. **Budget line items.** Brazilian legislation prohibits using resources budgeted for one purpose (e.g., electricity) for another (e.g., equipment retrofits/upgrades), something that is an essential element of ESPCs. Also, since energy and water services are classified as operating expenses, how best to classify the payments under an ESPC is unclear, as both investments and services occur over time. A related issue is the need to stipulate a line item within the RFP in the agency's annual budget to cover the expenses and assign a value, since those may not be known in an ESPC until M&V had taken place.

3. **Contract terms.** Brazilian law says the length of contracts should not exceed the time limit of the budget item that will pay for them (article 57 of Law 8666 of 1993). Since budgets are passed annually, this limit generally means that contracts must also be annual, though they can be renewed in successive years. ESPCs, which require longer durations to pay for improvements from savings and include pay based on performance, thus become problematic.

4. **Evaluation.** Brazilian law permits two basic categories of selection criteria: the lowest price only, and the lowest price and best technical proposal. This approach does not allow the proposal with the "highest benefit" to the contracting agency to be considered, although such a criterion is more relevant for ESPCs.

(continued)

Box 5.1 *(continued)*

These requirements have created barriers to the procurement of ESPCs in the public sector. As a consequence there have been only two known attempts at procurement using this mode: INFRAERO, the federal airport management company (RFP issued 1999, awarded 2000), and SABESP, Sao Paulo's water and sanitation utility (RFP issued 2005, awarded 2006). While both appear to be operating well, they are not yet replicable under current conditions.

Source: Poole 2008.

Figure 5.1   Schematic of Typical ESPC Procurement Steps and Key Issues

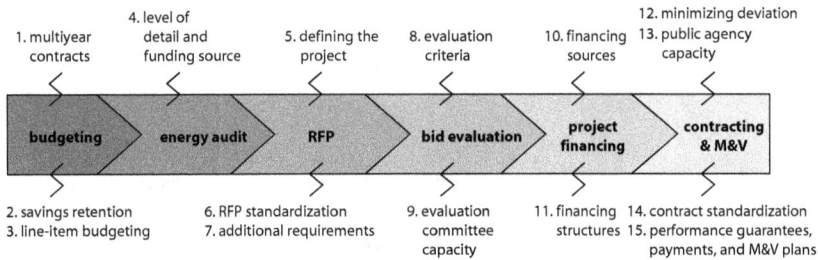

Source: Authors.

Initial energy audits
4. Level of detail and source of funds for initial audit

Development of the RFP
5. Definition of the project
6. Standardization of the RFP
7. Additional steps in the bidding process

Evaluation of bids
8. Evaluation criteria for multiple technical and financial parameters
9. Technical capacity of agency evaluating committees

Financing
10. Sources of financing
11. Financial structuring

Contracting and measurement and verification (M&V)
12. Minimization of deviation from the proposal
13. Capacity enhancement of public agencies
14. Standardization of contracting documents
15. Performance guarantees, payments, and M&V plans

These key six steps, along with discussion of their associated 15 issues, the approaches to them that different countries have taken, and some recommendations are addressed in the next chapter.

Note

1. For more information on typical procurement steps, World Bank guidelines, sample templates, etc., see http://go.worldbank.org/YZVQ9VQ490. A sample EOI for an ESPC RFP can be found in the IFC's *Manual for the Development of Municipal Energy Efficiency Projects* (India), 2008.

References

EVO (Efficiency Valuation Organization). 2007. *International Performance Measurement and Verification Protocol (IPMVP): Concepts and Options for Determining Energy and Water Savings,* Vol. 1, EVO-10000-1.2007.

Poole, Alan. 2008. "The Procurement of Energy Efficiency Performance Contracts in Brazil's Public Sector." Unpublished World Bank case study.

World Bank guidelines, sample templates, etc., see http://go.worldbank.org/YZVQ9VQ490. A sample EOI for an ESPC RFP can be found in the IFC's *Manual for the Development of Municipal Energy Efficiency Projects* (India), 2008.

CHAPTER 6

Key Public Procurement Issues and Options

As detailed in chapter 5, six key steps—along with 15 specific issues—have been identified based on the information elicited from government officials, experts, and practitioners in the international review and case studies that form the basis for this book. This chapter includes discussions of each of the main stages of the procurement process: budgeting, the energy audit, request for proposal (RFP) preparation, bid evaluation, financing, and contracting and measurement and verification (M&V). Each chapter section includes a discussion of the main issues for attention in each step, summarizes the various approaches used to address the issues, and offers some general recommendations.

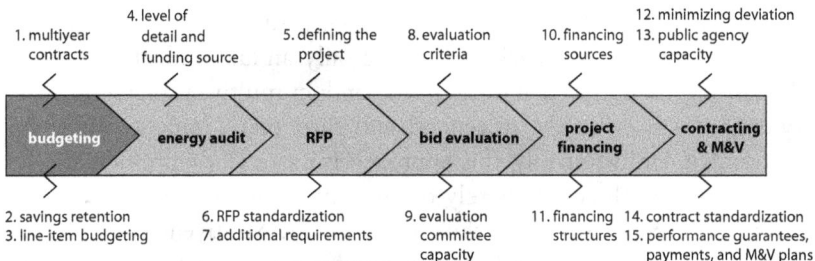

	4. level of detail and funding source	5. defining the project	8. evaluation criteria	10. financing sources	12. minimizing deviation
1. multiyear contracts					13. public agency capacity

| budgeting | energy audit | RFP | bid evaluation | project financing | contracting & M&V |

| 2. savings retention | 6. RFP standardization | | 9. evaluation committee capacity | 11. financing structures | 14. contract standardization |
| 3. line-item budgeting | 7. additional requirements | | | | 15. performance guarantees, payments, and M&V plans |

Budgeting

Before an energy efficiency project is typically considered, the public agency must identify the budget for the activity. Even though the project will ultimately pay for itself, upfront budget provisions are typically required to prepare the diagnosis or audit and assign staff to develop the bidding documents and supervise the project. Budget provisions are also needed if the agency wishes to finance the project on its own and ensure its ability to make future payments to the energy service provider (ESP) under the energy savings performance contract, or ESPC. Although the upfront budget requirements may not be large, the project will seldom occur if they cannot be secured. More important, typical budgeting rules and restrictions may make project financing and downstream payments more difficult and, if not addressed at the outset, could cause the project to fail before it is even initiated.

Dealing with the budget provisions for the upfront energy audit is covered in the next section; this section focuses on the issues associated with budgeting as it relates to the ESPC. Three critical budgeting issues emerged from the case studies and interviews:

Issue #1: How can a public agency enter into a multiyear ESPC if budgets are appropriated only annually?

Issue #2: How can a public facility retain energy savings benefits within its budget envelope, so as to pay the ESP and create an incentive for the agency to proceed with the energy efficiency project?

Issue #3: How can a public client deal with restrictions on line-item budgeting to pay for ESPCs? In other words, how can it finance capital improvements from operating cost savings, or pay the ESP from an existing budget line item (such as electric utility services)?

Multiyear Contracts

Because public sector budgets are typically submitted and approved each year, most public agencies receive only annual budget provisions. This arrangement makes it difficult to conduct multiyear planning, facility upgrades, and the like, in general, and particularly impedes multiyear contracting. Unfortunately, this approach is particularly problematic for ESPCs, since such contracts rely on multiple years of energy savings to permit the ESPs to recover their investments. Multiyear contracts are also crucial to ensure that ESP payments remain tied to performance; in other words, should the project performance decline in outer years, the public agency still has some recourse to the ESP.

Because many service contracts are required year after year, procuring them anew each year is not efficient. Therefore, many countries have found suitable ways of dealing with multiyear contracting. Multiyear contracts are required for many government transactions, such as public debt or hiring on open-ended contracts. But the typical annual budgeting framework does not always support them. There is a need to reconcile multiyear obligations under such contracts within the annual budget envelope, and tools such as the medium-term expenditure framework (MTEF), promoted by the World Bank and other donors, can help.[1] The MTEF can help ensure that commitments made by governments are consistent with the medium-term fiscal outlook. In countries that have adopted an MTEF, agencies generally may enter into multiyear energy contracts, as long as the contract is integrated with the existing budget process (included in any MTEF and as agreed with the ministry of finance), provided the budget planning process is sufficiently robust.

Some countries have also amended budgeting or procurement laws specifically to address ESPCs and the issues associated with multiyear contracting. For example, South Korea enacted a law in 1998 specifically to allow public entities to write multiyear ESP contracts beyond the fiscal year (Lee et al. 2003). Examples from Germany and the United States appear farther on in the chapter under Line-Item Budgeting.

As discussed in chapter 4, some countries have created utility-based ESPs, which can enter into ESPCs with public entities. They include Croatia (see Box 4.4, Utility ESPs in Croatia) and the United States (under the Federal Energy Management Program; see Case Study 1, in Part II). In many of these programs, it can be argued, the government agency already has an implicit multiyear service agreement with the energy utility, even without a formal contract, and therefore no contractual term restrictions should apply. Under this logic, utility ESPCs can be viewed as extensions of typical utility service provisions and thus not subject to typical public sector contract term limits. In Brazil, an option has been raised to characterize ESPCs as providing "continuous services," for which longer contract terms are permitted (up to 60 months). However, in such cases, the public agency may need to ensure that its future utility budget is increased if the project results in increased payment to the utility.

In cases where agencies judge that major changes in public budgeting are not practical in the near term, simpler models may need to be developed, such as one-year ESPCs, which may be better able to operate in the existing conditions (see Box 6.1 regarding one-year ESPCs in Mexico).

Box 6.1

Public Procurement in Mexico

Public procurement of ESPCs in Mexico poses a tremendous challenge. Most federal contracts, for example, can be issued for only one year, since agencies cannot obligate future budgets they do not yet have. ESPCs, which generally take more than a year for an ESP to recover costs, thus become largely infeasible despite model the enormous potential for cost and energy savings. In 2006, USAID developed a one-year ESPC to address this problem. Under this scheme, an ESP, once selected competitively, would design, procure, and install the project and then verify initial performance (at commissioning) to receive a payment of 60 to 70 percent of the contract value. A follow-on verification would take place three to six months later, to ensure performance has been maintained, after which the balance would be paid. In this way, if the project is financed by a third party, the government agency can still "pay from savings" as long as the debt service payments are less than the energy savings, even if the project's payback is more than one year. To allow for some recourse to the ESP in the event project performance substantially deteriorated in subsequent years, the public facility could also require an enhanced equipment warranty and/or a performance bond, perhaps around 10 percent of the contract amount, or similar provision for three to five years. Application of this model is now being explored by the government at the municipal level in cooperation with the World Bank.

Source: Authors.

Savings Retention

Because public agencies usually cannot carry balances from one budget year to the next, budget provisions are based on expected annual operational costs and outlays. In such situations, most public agencies actually have a perverse incentive to spend their full allocated budgets each year to help ensure that their subsequent annual budgets are not reduced. They generally have no incentive to invest funds in one fiscal year to save money in future ones, despite the benefit to the government as a whole. Energy savings projects are a good example of this dilemma; even though it is clearly in the public's interest to lower costs, an individual agency often has no incentive—even has a disincentive—to spend funds in one year to reduce energy costs in future years. Any investments in the current year reduce an agency's discretionary budget provision that year, and the future energy savings may result in smaller annual budgets in the future.

Aside from the public incentive issue is a contractual one. Because many ESPCs are based on paying the energy service provider from the energy savings, public agencies must be able to retain their energy savings in the current and future years to be able to make those payments.

One approach to the retention issue is to focus public ESPC programs initially on autonomous agencies, until the ESPC concept has been better tested and proven. In China, for example, local energy management companies (EMCs) have implemented many public sector projects over the past decade. However, these have primarily been with clients with substantial budgeting and procurement autonomy, such as water/heating/power utilities, hospitals, and schools and universities. EMCs have not yet been able to tap the vast potential for energy savings in other public agencies, such as federal ministries, which have more stringent budgeting and procurement regulations. Institutional agencies, such as schools and hospitals, have also had an easier time addressing energy savings retention issues in many countries. In Germany and Hungary, many budgets are either based on fixed costs or based on the number of students or patients; therefore, the entity has an incentive to reduce operating costs to free up budget space for other priorities, such as better service quality, new equipment, additional capacity, and so on.

In other cases, parent budgeting agencies that have more autonomy to retain operational cost savings can bundle and finance such projects among their subagencies. This bundling has been done on many occasions in school districts in the United States, where the district issues a blanket tender for all the schools and pays the energy service provider directly. Similarly, Thailand established an Energy Conservation Fund (ENCON Fund), which initially offered grants for public sector projects with the view that individual agencies may not have incentives to save energy on their own (World Bank 2001). Although the ENCON Fund financed the projects with grant funds, all the energy savings benefits ultimately accrued back to the ministry of finance (MOF) which originally capitalized the fund. (Ultimately, though, this public sector program was discontinued.) Brazil's private power distribution companies apply their demand-side management, or DSM, tariff surcharge, or "wire charge," to provide grant financing for public sector projects. Because these projects are grant funded, no payments back to the utility and thus no multiyear contracts are required. (Although this model does work in the existing system, it is not necessarily a fair, sustainable, or efficient model for the long term.)

A range of other approaches to the incentives and savings retention issues have been discussed, considered, or tried:

- Issuing a government mandate requiring public agencies to implement cost-effective measures, regardless of savings retention (China)
- Allowing the ESP to retain most or all of the energy savings for the full contract period, and then providing a noncash refund, such as a set of new computers, to the agency at the end of the project period (e.g., Austria, United States)
- Allowing the public agency to fully retain the energy savings only for the length of the ESPC (under consideration by the U.S. Federal Energy Management Program)

A summary of the various energy savings retention options, presented as a continuum, is summarized in Figure 6.1.

Line-Item Budgeting

Issues related to line-item budgeting are more complex. Many public agencies have two categories of budgeting: capital expenditures and operating costs. The categories are often not fungible, which makes ESPCs potentially problematic, inasmuch as capital improvements are essentially paid for from operating cost reductions under such contracts. For some entities, further complications arise because capital expenditures may be

Figure 6.1   Options to Address Savings Retention and Institutional Incentive Issues

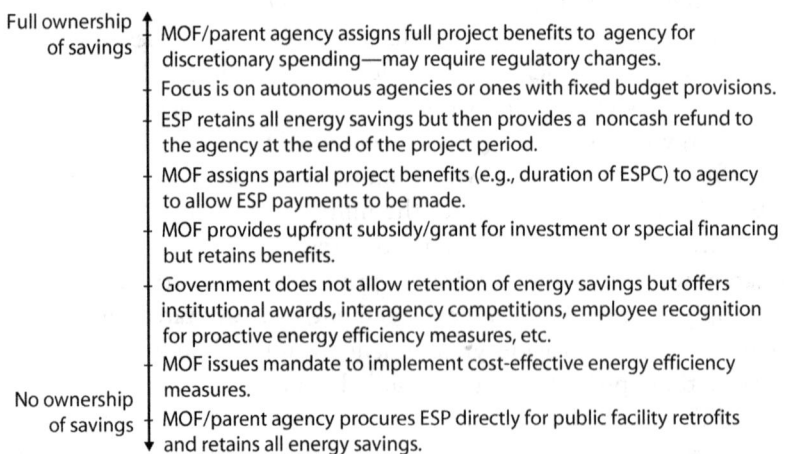

Source: Authors.

funded through central budget provisions, whereas operating budgets may derive from agency revenues, as is common in schools and hospitals. Other entities have even more restrictive line-item budgets that specify actual services, such as utility services, making it even harder for them to make payments to ESPs under ESPCs.

The United States has addressed a number of budgeting and procurement-related issues in a series of legislative and regulatory changes spanning some 25 years (see Box 6.2). Despite the strong government support,

Box 6.2

ESPC Budgeting in the United States

In the United States., many procurement and budgeting issues have been addressed through formal changes in legislation and follow-on regulations. The original ESPC legislation, the National Energy Conservation Policy Act (NECPA), was enacted in 1985. It directed federal agencies to improve energy management in their facilities and operations and gave them authority to enter into "shared energy savings contracts" with private sector ESPs. Legislation passed by the U.S. Congress in 1992 further authorized federal agencies to execute multiyear, guaranteed savings contracts and directed the U.S. Department of Energy (USDOE) to promulgate regulations for ESPCs. The Energy Policy Act of 1992, or EPACT (Public Law 102–486), amended NECPA to include additional energy management requirements and target reductions. Subsequent legislation and regulations relate to federal procurement of ESPCs.

As a result of this array of legislation and regulations regarding ESPCs and utility energy savings contracts, USDOE's Office of the Counsel and other agencies have provided guidance and legal opinions regarding interpretation of the intent and implementation of the various regulations. For example, agencies can retain 50 percent of the energy and water cost savings from appropriated funds for additional energy projects (expected to increase to 100 percent), including employee incentive programs. However, with continued pressure to limit federal agency budgets, any energy cost savings that remain after repayment of the ESPCs are often being immediately cut from agencies' operating budgets. Retention of energy savings will continue to be a concern of the agencies during the current period of budget shortfalls, as budget examiners continually look for ways to reduce operating expenses and even possibly try to cut agency budgets before full repayment of the ESPC.

Source: Case Study 1.

however, elements of ESPCs continue to be reviewed and discussed, so that ESPC procurement and budgeting procedures in the United States will likely continue to evolve.

In Germany, federal properties entering into multiyear ESPCs with total annual payments to the contractor (out of energy savings) of more than €300,000, or in which the investment element is considerably greater than the service component, require a commitment appropriation (that is, an authorization for financial commitments beyond the current fiscal year) in the federal budget for the future payments to the contractor. That commitment generally must occur before the tender process begins. At the level of the states, there is more uncertainty for longer-term ESPCs, because supervisory authorities in different German states have different views and diverging practices on how to handle them. The most liberal state regarding ESPCs is North Rhine–Westphalia, where local authorities do not need a separate authorization for each ESPC. In six other states, each ESPC has to be authorized separately, and its value is usually counted against the debt limit of the municipality. The remaining states fall somewhere between those extremes. In the city-states, such as Berlin, Bremen, and Hamburg, there is no separate supervising authority.[2] The PICO model used in Stuttgart is another way to address these issues, in which an internal service provider receives a type of revolving fund from which to make many of the energy savings investments and then recovers the investment through budget transfers under a shared savings arrangement.

Issues related to budgeting for ESPCs in India are still being explored. At present, capital expenditures of many public agencies are paid for from central budgets, and operating costs are generally supported from separate revenue budgets. For the initial central government ESPCs with the Indian Central Public Works Department, payments were made from operating cost budgets as "planned" expenditures. However, the municipal contracts in Tamil Nadu and Gujarat are being supported through specific government orders, which direct local bodies to implement ESPCs and pay the energy savings from the revenue budgets at the state level. In this way, the budget allocations for subsequent years are not necessarily reduced, since the current year's energy costs are met through the revenue budgets.

Recommendations

Public budgeting systems follow similar principles, but the interpretation and application can be quite different, even among different agencies within one country. Broader budgeting issues and decisions typically require higher-level engagement and discussion, which can be difficult and lengthy. Countries are not advised to seek sweeping changes to budgeting laws without actual experience regarding how ESPCs will work best under local conditions and how local ESP markets will develop. If one is to avoid the risk of adopting ill-suited regulations, it is recommended to actively seek out simpler options in the structuring of the ESPC, to seek special waivers for pilot projects, or to reach some basic agreements with parent budgeting agencies until the ESPC concept can be tested and its benefits demonstrated. As public budgeting systems mature, it is expected that some of the budgeting issues associated with ESPCs, such as the multiyear contracts, will be solved. Further, with increasing decentralization of budgeting to local levels and more decision-making autonomy, public agencies, particularly at the municipal level, may have more authority to adjust the usual procurement and budgeting restrictions to test new ways to lower operating costs, such as ESPCs. Clarification on budgeting issues may be needed with respect to ESPCs over the longer term. Budgeting issues likely will continue to hamper the adoption of ESPCs in many developing countries in the near term. In such cases, the following are recommended:

1. Starting public procurement schemes with more autonomous public entities, such as municipalities, universities, hospitals, and water utilities, to test and refine the ESPC concept. As more experiences are gained and demand throughout the government sector increases, increased pressures to find suitable longer-term budgetary solutions will arise, and precedents will emerge.

2. Selling the concept of ESPCs—that is, reducing operating costs without upfront investment capital and paying for the services from savings—directly to parent budgeting agencies to gain their support. This may require some upfront audits to identify high-return projects that will convince these agencies to join the effort to find solutions, so that these benefits can be realized. The parent agency could be the school district, municipality, provincial department, line ministry, or the ministry of finance or treasury. The parent agency can provide key input on how best to address the budget issues, to allow the project to move forward, even if only on a pilot basis initially.

3. Creating a public energy efficiency financing program (see the Financing section in this chapter) after a few projects have been implemented, to

deal with some of the incentive and financing issues. With the development of such a program, questions related to how projects are financed and how public agencies pay the ESP can be dealt with in the program design, possibly allowing for some of the more stringent budgetary obligations to be waived.

4. Initiating changes to the budgeting laws and supporting regulations in the longer term. For such situations, it is much better to have actual ESPC examples to bolster the case for amending laws, with documentation of benefits that existing public agencies have achieved, to help make the case. Successful past projects allow precedents to be established.

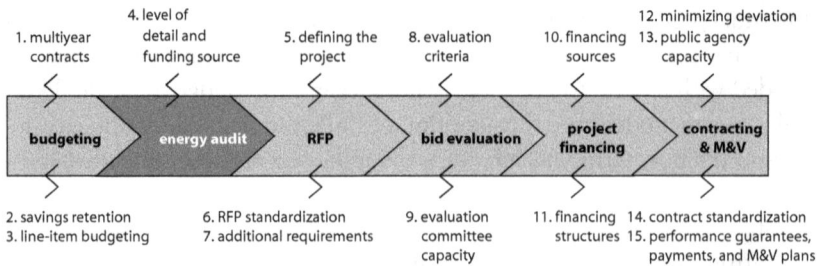

1. multiyear contracts
4. level of detail and funding source
5. defining the project
8. evaluation criteria
10. financing sources
12. minimizing deviation
13. public agency capacity

budgeting | energy audit | RFP | bid evaluation | project financing | contracting & M&V

2. savings retention
3. line-item budgeting
6. RFP standardization
7. additional requirements
9. evaluation committee capacity
11. financing structures
14. contract standardization
15. performance guarantees, payments, and M&V plans

The Energy Audit

The initial energy audit—which is also called the upfront diagnostic, prefeasibility study or the baseline assessment—is an inspection of a building or facility to assess current energy consumption, to identify methods for reducing it, and to evaluate the potential of various energy cost saving measures. This inspection process identifies changes to a building's structure, equipment, systems, and/or operational procedures that will result in a reduction of energy consumption and provide sufficient energy cost savings to pay for the improvements over a reasonable number of years.

The energy audit is an important first step to assess how much energy a public facility consumes, determine whether or not cost-effective energy efficiency options exist, and identify which systems merit more detailed study. The audit is integral to an ESPC, because the agency must confirm that cost-effective energy savings opportunities exist, must assemble credible technical data and analysis to reduce the cost to bidders of preparing their proposals, and must provide notional definitions for the project scope that will be part of the RFP. It is a simple but important hurdle in getting the project to the next step. The issue for public agencies is as follows:

Issue #4: What is the appropriate level of effort for the initial audit? Who pays the cost, and what types of qualified entities should conduct it?

Level of Detail and Funding Source

Types of energy audits. The term "energy audit" can be quite subjective and vary from country to country. Thus it is important to realize that certain types of energy audits have different connotations, especially for the individuals that work with buildings, such as the maintenance staff, the facility energy manager, the equipment vendors, the engineering firms, and the other ESPs. Energy audits can be conducted in varying degrees or levels of technical detail, accuracy, and complexity based on the purpose. Energy audits have generally been found to fall into three basic categories—walk-through, preliminary, and investment grade (see Table 6.1)—although a number of countries call them by different names.

Table 6.1   Types of Energy Audits

Types of energy audits	Description
Walk-Through Survey (also called a screening audit or initial diagnosis)	Brief on-site inspection of a facility to evaluate the potential for energy cost savings measures, gather information to determine the need for a more detailed audit, become familiar with the building operation, and develop limited detail on corrective measures.
Preliminary Energy Audit (also called a preliminary assessment)	Evaluation of the energy unit and cost savings potential, building conditions, energy consuming equipment, and hours of use or occupancy, for the purpose of developing preliminary technical information prior to issuance of the request for proposal.
Investment Grade Audit (also called a detailed feasibility study or detailed project report)	Detailed analysis of the energy cost savings and energy unit savings potential, building conditions, energy consumption, and hours of use or occupancy for a facility, for the purpose of preparing final technical and price proposals that could lead to an ESPC. Also includes economic and cash flow analyses.

Source: Authors.

For the purpose of ESPC procurement, all three are relevant. In many countries, such as Canada and India, a *walk-through survey* is typically performed by a third party engaged by the public agency, and the results included in the RFP. A third party is often required because conflicts of interest can arise if the firm that conducts the energy audit is also allowed to bid on the resulting project. During the bidding process, each bidder is invited to visit the host facility and conduct its own *preliminary energy audit*, so as to be able to prepare proper technical and financial proposals. Once the contract is awarded, the selected ESP conducts an *investment grade audit, or IGA*, to finalize the precise details of the ESPC, document the baseline, and develop the detailed project design. For indefinite quantity contracts, or IQCs, the ESP would typically do the walk-through survey and, if suitable measures are found, proceed directly to the IGA.

The rationale for the seemingly inefficient three-tiered auditing process (i.e., the initial survey conducted by the public agency, the more detailed audit conducted by ESP bidders, and the IGA conducted as part of the contract) has to do with the technical credibility of a third party audit, the need to ensure competition by lowering bidding preparation costs, and the allocation of risk. While walk-through surveys are rather straightforward and would generally look similar even if carried out by different firms, detailed audits would have more variation. This conclusion is to be expected, since firms will have different levels of technical expertise on each energy system, varying access to proprietary information on certain new technologies, different experiences from which to make certain estimates, and so on. As a result, more detailed upfront audits by public agencies will produce diminishing returns. ESPs that are bidding against an RFP with a more detailed energy audit would likely conduct their own analysis anyway to confirm the data and savings opportunities, although this would increase the bidding costs. Relying too much on another firm's detailed audit is also very risky because the bidding ESP is ultimately responsible for achieving the energy savings presented in its final bid. Overly detailed audits performed upfront may also lead to a more prescriptive project definition in the RFP, which could inhibit innovations by some of the bidding ESPs.

In terms of risk allocation, each energy audit is associated with a cost. The walk-through audits are a simple way to provide basic technical information that can easily be verified, including an inventory of equipment, billing information, facility function and age, and so forth, to help lower bid preparation costs for the ESPs. More detailed audits, of course, require greater effort and thus incur higher costs. Bidders that are required to conduct upfront, detailed energy audits during bid preparation run the risk that they will not be compensated for the costs if they are not selected. Therefore, the more onerous the RFP requirements are in terms of level of detail of the audit and proposal, the fewer bidders there are likely to be, which can reduce the level of competition. At the same time, if the RFP does not require sufficient detailed analysis, it creates possibilities for bidders to "over-promise" so as to present a high level of energy savings in the proposal and win the contract. Without sufficient technical information at the outset, agencies are unlikely to know what level of energy savings is reasonable and thus may have greater difficulty assessing the credibility of some proposals. Agencies thus need to balance the level of detail of the initial energy audit with the costs and risks.

Even if a public agency is committed to implementing energy efficiency projects using an ESPC, it may lack the internal expertise to conduct a quality energy audit or the funds to hire a qualified firm to conduct the audit on its behalf. In addition, public facilities rarely have submetering to identify how energy is being used in different parts of the facility, and the limited facility staff members are often unable to diagnose and respond to energy saving opportunities. Trying to identify qualified energy auditors can also be difficult. Often such auditors lack sufficient practical experience to properly assess the mechanical and electrical systems of buildings, or they do not provide meaningful and practical recommendations in the energy audit report.

Options for funding the initial energy audit. The international review found that countries have approached the energy audit issue in a range of ways. In most cases, the audits were performed by public agency technical staff, utilities, engineering firms, universities, or other ESPs:

- *Mandatory energy audits for public buildings.* A number of countries, such as China, the Czech Republic, India, Thailand, and Tunisia, have required some form of mandatory energy audit for certain public facilities to identify cost-effective energy savings projects. Although this requirement has had positive impacts in some European countries, the experience with mandated audits in developing countries has been quite poor. Such requirements have often resulted in agencies' securing low-cost audits to meet the requirements, often to the detriment of the quality of the final report, and have rarely resulted in meaningful investments and energy savings.

- *Free or low-cost energy audits by public entities.* In China and Vietnam, public entities such as public utilities or provincial/regional energy conservation centers can offer energy auditing services to public agencies, often free or at a subsidized cost. Other countries, such as Brazil, Canada, and Thailand, have negotiated with local utilities to supply preliminary energy audits under utility demand-side management (DSM) programs, which are often funded by a DSM surcharge on their consumers' electric bills. These preliminary energy audits are often free to public agencies, but some programs threaten to impose cost penalties if the recommendations of the audit are not implemented within a specific period that the utilities establish.

- *IQC selection of ESPs.* As explained in chapter 4, indefinite quantity contracts, or IQCs, involve a competitive selection process, typically at the national level. Selected ESPs submit proposals and can enter into direct contracts with certain public agencies without the need for further competition. This approach removes the risk to the ESP that it will not be compensated for any initial energy audit or that, if the audit is approved, it may not be selected to implement the project. Both Hungary and the United States have used this approach. If the public entity decides not to proceed with the project after the detailed audit has been completed, the agency is usually required to compensate the ESP for the audit.

- *Representative energy audits.* For campus settings at colleges or other large public facilities with many buildings, or for a bundle of similar government office buildings, the preliminary energy audit may only involve a subset of the target buildings to minimize auditing costs. Care must be taken to select these buildings appropriately and extrapolate the resulting information across all the facilities. A number of countries, such as Canada, and individual states in the United States, such as New York and Illinois, have used this approach.

- *Energy audit templates.* To assist agencies in the energy audit process and streamline data collection efforts, a number of countries, such as Canada[3] and the United States,[4] have created guidelines for energy audits that include sample forms, worksheets, and graphs to be used for data collection and as deliverables at specific audit levels. This guidance can help agencies complete their own audits or help organize data for auditing firms or ESPs.

- *Compensation for all ESPs.* Another way to deal with the audit costs and risks is for the public agency to agree to compensate all ESP bidders for some or all of the cost of their detailed energy audits as part of their bid preparation. This approach is sometimes used in Canada and France, where a public authority may compensate the bidders for a portion of the cost of their technical and financial studies. Although it increases costs and can create some perverse incentives for ESPs, this approach is one way to require more detailed audits during bid preparation without reducing the likelihood of a strong pool of ESP bidders.

As noted earlier, there is clearly a wide range of options for dealing with the initial audit. As can be seen in Figure 6.2, these approaches range from more prescriptive energy audits that may define the energy efficiency project, to more flexible requirements for basic technical information. Each option has implications with respect to who should conduct the audit and how much it will cost. Which option is best depends on several factors, such as the complexity of the host facility's energy systems, the available budgetary resources for an external audit, the technical expertise of the facility staff, the specificity of the project definition in the RFP (see the section Request for Proposals: The Bidding Documents), and so on. Ultimately, the process should be as flexible as possible and driven by the client's needs.

Figure 6.2   Illustrative Options for Initial Energy Audit

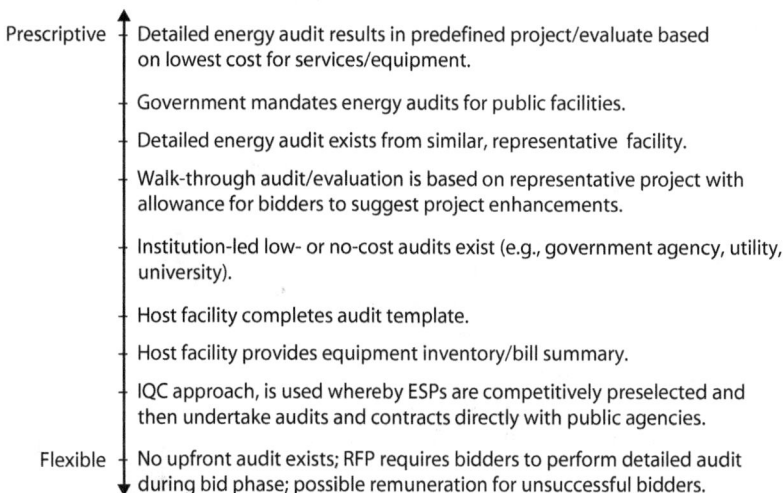

Prescriptive — Detailed energy audit results in predefined project/evaluate based on lowest cost for services/equipment.

Government mandates energy audits for public facilities.

Detailed energy audit exists from similar, representative facility.

Walk-through audit/evaluation is based on representative project with allowance for bidders to suggest project enhancements.

Institution-led low- or no-cost audits exist (e.g., government agency, utility, university).

Host facility completes audit template.

Host facility provides equipment inventory/bill summary.

IQC approach, is used whereby ESPs are competitively preselected and then undertake audits and contracts directly with public agencies.

Flexible — No upfront audit exists; RFP requires bidders to perform detailed audit during bid phase; possible remuneration for unsuccessful bidders.

Source: Authors.

Recommendations

Energy audits can be provided in varying degrees of detail (or technical levels), and each option has implications for the project definition, data requirements, level of effort, cost, and level of required expertise. Recent efforts have shown a movement away from broad-scale mandates for energy audits toward a focus on when an energy audit should be performed, by whom, and at what level of detail.

Preliminary conclusions suggest that it may be possible to replace the typical upfront energy audit requirement with basic technical data provided by the public agency. Often the necessary technical information can be prepared at a very low or no cost by the agency staff members themselves. The data would include the following:

- A description of the buildings, their functional use (office space, cafeteria, training center), age, hours of use or occupancy, conditioned square footage, meters, and other relevant building characteristics
- At least 24 consecutive months of recent utility bills, including consumption, demand, cost/tariff information, other fuel costs, water bills (if water savings projects are being considered), and any other relevant energy consumption data
- An inventory of all major energy-using equipment, its age, capacity, hours of operation, and rated capacity data
- Any available information on corrective measures, target systems, and past energy efficiency measures implemented
- Any additional information that may help bidders identify opportunities for energy cost savings and determine the feasibility of an energy efficiency project

Bidding ESPs can augment these substantive data with additional information collected during their site visits in the bid preparation process.

In countries where a third-party energy audit is deemed necessary, or in facilities with more complex energy consumption such as water utilities, the level of detail of the energy audit should be carefully considered. It may even be advisable to hold upstream consultations with prospective bidders to discuss technical energy consumption information requirements and gauge the level of detail they require, as well as what data they would prefer to collect on their own. Some countries may have procurement rules that require more prescriptive RFPs, which will then require greater detail in the upfront audit. Each country has its own unique procurement rules, and each level of government and even each agency may have its own preferences, which should be taken into account when determining the audit requirements.

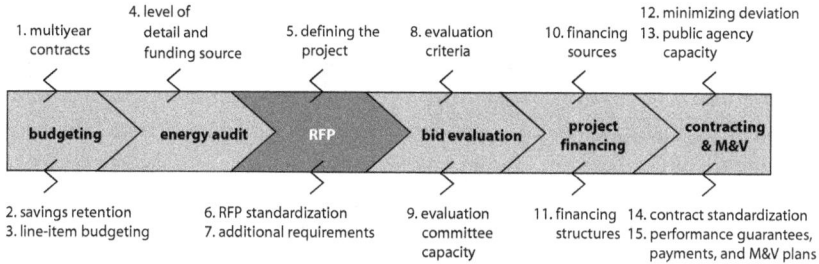

Request for Proposals: The Bidding Documents

The selection by public agencies of a competent ESP to provide energy efficiency services using an ESPC is generally made through a competitive bidding process, which involves the development and issuance of a Request for Proposals (RFP), also referred to as "tender," "solicitation," or "bidding documents." The success of the resulting energy efficiency project depends on the effectiveness of the competitive procurement, which can be significantly enhanced with a well-developed RFP. The complexities of ESPCs, which involve the integration of technical, economic, financial, and operational elements, influence the design of the bidding documents used in the procurement process. This section discusses the following issues related to the bidding documents used for competitive public procurement of energy efficiency services:

Issue #5: How should the energy savings project be defined in the RFP to allow for reasonable comparisons but also innovative proposals? According to this project definition, which RFP type should be used—goods, works, services, or a combination?

Issue #6: When should the public ESPC procurement process be standardized? Should standard RFP documents and contracts be developed?

Issue #7: What additional steps should be included in the RFP, such as prequalification, upstream consultations and pre-bidding conferences, facility site visits, oral presentations, and so on?

Defining the Project

A basic issue in the procurement of energy efficiency services is what is actually being procured by the public agency using the ESPC approach. This is critical, as many countries, the World Bank, and other donors have different procurement methods and have standardized documents depending on the procurement category. In practice, ESPCs are combinations of goods, works, and services. Services include IGAs, project design, measurement and verification (M&V), operations and maintenance (O&M), training, and more. Goods include procurement (and sometimes installation) of products and equipment. Works may involve revamping existing systems, constructing stand-by power and cogeneration units, and the like.[5] The specific details of the bidding documents (and the resulting contracts) vary with respect to the terms and conditions pertaining to the actual procurement type. When the project is defined in the RFP, three issues must be addressed: (a) specifying the type of procurement (goods, works, services, or a combination); (b) defining the project objectives and basic parameters; and (c) defining the project's scope of work.

As Table 6.2 shows, different countries and organizations view ESPCs differently. In the United States, the Federal Energy Management Program created a special contract and procurement method just for ESPCs, recognizing that they did not properly fit any of the existing methods. France has also created a specialized category under its Public-Private Partnership Law. Germany generally uses works contracts (referred to as "VOB") for federal properties, but in other jurisdictions uses service contracts (called "VOL") on the premise that the delivery of energy savings is more of a service. Many of the countries that have used utilities and public ESPs to subcontract with local firms to do the work also generally classify their ESPCs as service contracts. There is no one right approach; each depends on existing regulations and how ESPCs are interpreted (and, therefore, classified) within them.

Table 6.2   Scope of Procurement in the Request for Proposal (RFP) Document

Country/Institution	Type of procurement
India (Tamil Nadu)	Goods and services
India (Gujarat)	Works and services
Germany	Works or services
United States (NYPA)	Services
United States (FEMP)	New law/procedures
France	New PPP law/procedures
World Bank	Management contract (goods, works, and services)

Source: Authors.

The World Bank does not have standardized documents to support ESPCs, and until recently it had no approved template for combining goods, works, and services in one contract. In December 2007, the World Bank approved a standard RFP template for "Management Services,"[6] which includes provision of goods, works, and services and allows for performance-based remuneration, as well as other features (e.g., output-based rather than input-based requirements) typically contained in ESPCs. This document now creates strong precedent for allowing the World Bank to increase the use of ESPCs in its lending programs. However, a guidance note, along with a template "Terms of Reference," will likely be needed to advise clients on how to use the Management Services template for energy efficiency projects.

The proper definition of the "project" is another important aspect of the ESPC process. Establishing the project objectives and basic parameters will guide the procurement type, how and which ESPs will bid, and how the bids will be evaluated. Although the level of specificity of the energy efficiency measures to be implemented is determined by the bidders, public procurement rules and regulations may require the public agency to establish some key boundaries that define the project in the RFP, to ensure transparency in the subsequent evaluation of the bids without prescribing the manner in which the project should be implemented. Some countries also require that the budget be specified upfront, or that an independent government cost estimate be developed before the RFP is issued. Most countries, as well as the World Bank, would not allow an RFP to be issued that simply invited bidders to propose their best energy savings project for a given facility without any basic parameters. Such a process would be viewed as too vague, may invite ESPs to propose improvements beyond the intent of the host facility, and may limit the ability of the tendering agency to gauge the relative merits and cost efficiency of the proposals submitted. The challenge, therefore, is to set output-based parameters to accommodate the evaluation of a variety of technical approaches and solutions. Bidding documents must thus provide very clear project objectives, which various approaches will be measured against, as well as lay out a clear methodology for evaluating dissimilar bids.

Even then, defining the project is not straightforward. In some countries, public agencies rely on their preliminary energy audit to prescribe the specific energy efficiency measures to be installed. This is the most restrictive case, but it is not uncommon (see Box 6.3 for an ESPC example from the Arab Republic of Egypt). Brazil's procurement legislation requires that a detailed, prescriptive project description be used, so that bidders can be

Box 6.3

Lighting Efficiency Improvements in the Arab Republic of Egypt

In 2005, the Ministry of Water Resources and Irrigation (MWRI) of the Arab Republic of Egypt developed an RFP that focused on energy efficiency improvements in the lighting systems in its headquarter building in Cairo. An energy audit was conducted and shared with bidders. But the procurement rules required the project to be predefined in the RFP, that is, the target systems and the technical specifications of equipment to be used had to be identified in the RFP document. The rationale was that allowing for different bids would make the evaluation complex, could ultimately lead to the agency's paying too much for the project, and could lead to challenges by unsuccessful bidders. Although the resulting cost proposals were easier to assess, the scope for bidders to offer improved designs and innovations was severely limited. The RFP included a clause allowing bidders to propose enhancements to the project design, but they had first to win the bid (on both technical and financial terms) based on the given design. The RFP incorporated some performance clauses that were compatible with Egyptian procurement regulations, such as that a portion of the payment would be withheld until a third-party M&V report was prepared, along with a 5 percent bond, but the contract duration was less than a year.

Some ESPs that had expressed interest at the pre-bidding conference chose not to submit proposals because of the limited scope for innovation and the fact that more traditional ESCOs would not be able to compete with equipment suppliers on price. Nevertheless, three proposals were received, and an award was made in 2007 for about US$84,000. The project is now performing well, saving MWRI about US$30,000 a year.

Sources: Nexant 2008; MWRI 2007.

assessed based on the same project. If the agency decides not to prescribe the energy savings measures in the RFP, it may be able to get a wider range of innovative proposals, but the comparative evaluation of the bids becomes more difficult (see Evaluation section later in this chapter). A compromise approach may be to prescribe the systems to be retrofitted but not prescribe the technical solutions to be used. Alternatively, the agency could be flexible on the energy systems targeted, but prescribe some minimum set of energy efficiency measures that all bidders must include in their proposals, leaving bidders the option to provide additional measures as they deem appropriate. The evaluation and comparison of the bids can then be conducted based on the mandatory measures, but the selected ESP would be allowed to negotiate additional or alternative ones as well.

Another option for defining the project is to specify a minimum level of energy savings. Some recent RFPs in India have required that the ESP achieve a minimum percentage of savings. Such a requirement is not as common in more mature markets for public procurement, such as in Canada, Japan, and the United States, but in underdeveloped markets it can be a good way to offer a comparable parameter for the project and avoid any concerns over "cream skimming." Germany and India have both used minimum energy savings as a means of defining projects (see Box 6.4 for an example from India).

Box 6.4

Minimum Savings Requirement in India

The Tamil Nadu Urban Development Fund (TNUDF) was established in 1996 as an autonomous financial intermediary incorporated as a trust fund with private equity participation. TNUDF is charged with financing urban infrastructure and poverty alleviation projects while facilitating private financing of infrastructure. In 2007, TNUDF initiated a project to support ESPCs in municipalities based on a 2005 World Bank Public-Private Infrastructure Advisory Facility (PPIAF) report that presented a framework for private sector participation in municipal energy efficiency programs and defined the application of the ESPC approach.

The TNUDF issued an RFP for urban local bodies (ULBs) to local ESPs. To prevent cream-skimming, it required that the ESP must achieve minimum savings of 30 percent in each of its proposed projects in municipal pumping and street lighting. If the IGA conducted by the ESP indicated lesser savings, not only was the project terminated, but also the IGA costs were not reimbursed by the public ULB. The ESPs were encouraged to contact and visit the ULBs (at their own expense) before submitting proposals, so ESPs could ensure that the savings were achievable. This stringent requirement of 30 percent minimum required savings may have discouraged some short-listed firms from bidding on the project. For those ESPs that did submit a bid, the significant incentive to estimate at least 30 percent savings was significant. But it is also important that the minimum level be achievable.

(continued)

Box 6.4 *(continued)*

In the TNUDF case, it appears that achieving the 30 percent target may be diffi-cult, which could threaten the project. A more recent RFP issued in the state of Gujarat in India had a similar requirement but with a smaller minimum of 20 percent savings (*"ESP shall guarantee at least 20 percent of the savings over the existing energy bills. Otherwise, the bid will be considered as non-responsive"*).

Sources: Tamil Nadu Urban Infrastructure Financial Services Ltd. 2008; GUDC, 2008.

A related option is the specification of a minimum share of the sav-ings to be received by the public agency. Most countries use RFPs that give bidding ESPs flexibility to propose the share of energy savings to be allocated to the public host facility. The proposed sharing of savings is an important criterion in evaluation and in the selection of the win-ning bidder. However, some countries have chosen to specify in the RFP a minimum share of the savings to be given to the public agency (with the ESP providing 100 percent of the financing). The Gujarat RFP in India contained the following requirement: "The ESP shall share 20 per-cent of the savings with the municipality. Otherwise, the bid will be considered as non-responsive" (GUDC 2008). Similarly, two projects in Brazil specified allocation of a minimum of 10 percent of savings to the host facility.

It is unclear whether specifying the minimum share of the savings pro-vides an advantage in the bidding process. On the one hand, it assures the public agency of a guaranteed minimum monetary savings and helps reduce variation in the bids submitted. On the other hand, it may lead the bidders to offer no more than the minimum specified share to the agency, perhaps limiting the agency's project benefits. Allocating a bigger share of savings to the client can also result in longer contract durations and thus higher financing costs. In Canada, many of the public ESPCs use the "first-out" method, in which the ESP takes 100 percent of the savings until it recovers its investment with fees. This approach encourages greater ESP competition in the bidding. Creating a minimum energy sav-ings share to the facility would not allow first-out ESPCs and might also be incompatible with some of the other ESP business models.

Finally, the project definition must include the package of services that the ESP must provide to the public host facility. As noted earlier, not all of the services typical of Western-style energy service companies, or ESCOs, need to be included if the agency does not want them or if likely bidders are unable or unwilling to provide them. And not only ESCOs may bid, which is why this book generally uses the term "ESP." Most ESPCs would include detailed audits/IGAs, engineering and project design, procurement, installation, commissioning and M&V; others may also require project financing, performance guarantees, O&M, and training. Some agencies may prefer to leave some of the tasks vague, to see what packages of services the ESPs may offer on their own. As with other aspects of the ESPC, this arrangement can lead to some innovation but can also make evaluation of the bids more difficult. Public agencies also need to consider that the more services they request and the more risks they assign to the ESP, the higher the costs. If additional costs must be incurred, such as program reporting to a central agency or donor, or if additional revenues, such as carbon financing, will be available, the RFP must be clear with respect to which of the parties will bear the costs and own the benefits. And whereas outsourcing the entire suite of functions can be appealing to the public entity, ESPs are likely to have different strengths and may not be equally qualified to conduct all of the tasks envisaged.

RFP Standardization

The development of a standard RFP package can help public agencies save time and cost in the early stages of the procurement process. Both the U.S. Federal Energy Management Program (FEMP)[7] and the Canadian Federal Buildings Initiative (FBI)[8] have developed and used standard bidding documents and procedures. Also in the United States, the Energy Services Coalition[9] has developed standard RFP documents for use by its members.[10] The Clinton Foundation, through its Large Cities Climate Leadership Program (also known as the "C40 program") has attempted to develop a standard RFP that is applicable across many countries, and that work is ongoing.[11] In India, efforts have been devoted to developing standard RFP documents for public buildings[12] and for municipal pumping and street lighting projects.[13]

A review of some of these standard documents reveals substantial differences from one country to another, but the general content is similar and can be adapted to the situation in any country. The advantage of standardizing the documents is improved efficiency in the procurement process, reducing transaction costs of both public agencies and bidding ESPs. It should be noted that the RFP documents of countries with substantial ESPC experience have evolved over time. Standardization was achieved only after considerable experience was gained from earlier procurements. Development of standardized templates too early in this evolution can inhibit innovation and risks institutionalizing suboptimal approaches. Prescribing a standard RFP may also impose constraints on the procuring agency and on potential bidders and limit their ability to be flexible and innovative.

Additional Requirements

Prequalification of ESPs. Some countries or agencies include a prequalification or short-listing step in the procurement process, issuing an invitation for prequalification or request for expressions of interest (EOI) to a large number of ESPs using statements of qualifications that interested ESPs submit, the agency then develops a prequalified or short list of ESPs that will be invited to respond to the RFP.[14] The purpose of prequalification or short-listing is to ensure that only fully qualified and capable firms are invited to submit proposals. Short-listing is generally a quicker process, and for that reason it is attractive in some cases. In particular, short-listing may make sense if the contracting agency already knows that the pool of eligible bidders will be limited to some number not exceeding six. Otherwise, prequalification is generally preferable, but it requires more work to determine the criteria for prequalification. For countries using the IQC or public ESP approaches, such a screening process for bidders is not typically used. However, the new U.S. FEMP IQC has 16 selected ESCOs, so that agencies seeking competitive solicitations under that contract will likely wish to short-list two to five firms. This step has been used in most of the projects in India, as the process is still new and the participating municipalities wanted to prevent low-quality proposals.[15] However, additional screening steps will likely increase the bidding period, and short-listing can reduce the level of competition. Where it is not used, the agency can directly proceed to the issuance of the RFP.

Conducting the IGA. The negotiation and execution of an ESPC between the public agency and the ESP requires an investment grade audit, or IGA, that documents in detail the energy efficiency measures to be installed. Bidding documents take various approaches with respect to IGAs:

1. The public agency has a third-party auditor complete the IGA and includes it in the RFP;

2. The RFP requires each bidder to conduct an IGA as part of its bid preparation; or
3. The RFP requires the bidders to conduct the IGA as a part of the contract, in which case the agency may provide a preliminary audit or some facility characteristics information in the bidding documents.

An advantage of the first approach is that the agency obtains a good idea of the types of energy savings measures and the amount of potential savings before it issues the RFP, and all bidders have the same information with which to work. It can also reduce the cost of bid preparation, potentially expanding the pool of bidders. A disadvantage is the expense to the public agency. Additionally, bidding ESPs would consider it risky to prepare a proposal based on technical information gathered by another firm. Given these risks, most bidders would likely choose to conduct their own audit anyway, making the initial IGA redundant and a waste of money. Many ESPs also believe that their audits can lead them to identify innovative technical solutions to improve the project energy cost savings.

The second approach, requiring bidders to conduct the detailed IGA at their own expense as a part of their response to the RFP, has been used in some cases but is not preferred because the cost burden on the bidders is likely to limit the number of responses. (This is why some countries, such as Canada and France, have provisions to partially compensate ESPs for bid preparation costs even if they are not selected.) Unfortunately, without the IGA, the details of the ESP's financial proposals may be subject to minor adjustments after the contract award has been made and the IGA is then conducted, which may or may not be acceptable to some procurement officers.

The more mature public sector performance contracting programs, in Canada, Germany, and the United States, for example, use the third approach and require the ESP to conduct the IGA in the first phase of the contract, and then, after a review of the audit findings, to negotiate the performance contract.[16] This approach appears to be becoming more generally accepted in other countries. In India, the Delhi water utility (Delhi Jal Board or DJB) initially attempted to adopt the first approach and had an investment grade audit completed by a nongovernmental organization (NGO). However, it became clear during the pre-bidding conference that ESPs would not accept this audit and would have to do their own. The DJB, therefore, changed the approach and revised the bidding documents to adopt the two-phase contracting approach (SRC Global Inc. 2005). In the province of Quebec, Canada, a public agency may conduct its own audit, sometimes with support from utility grant programs, and then make the results available to potential ESP bidders. But the trend is moving away from conducting audits prior to the RFP process.

Pre-bidding conference. In view of the complexities of the ESPC approach and the numerous elements of the bidding documents, most RFPs include a pre-bidding (or pre-bid) conference with the potential bidders. Such a conference provides an opportunity for the bidders to have their questions and concerns addressed and to point out any items that may require modifications to achieve a smooth and effective procurement process. Some countries make attendance mandatory to ensure that all bidders have a common understanding of the RFP requirements; others make it optional.

Another question is when to hold the pre-bid conference, for example, before the RFP is finalized or after it has been distributed to the prequalified or short-listed firms.[17] In less-developed markets, early consultation with potential bidders is very important. It may be worth holding upstream meetings even before the RFP is developed, to gauge ESP perspectives about how much technical information they would prefer to see in the RFP; what types of services and risks they would be willing to accept; and preferred provisions regarding contract duration, payment schemes, performance guarantees, and so on. Similarly, another consultation once a draft RFP has been developed may also be worthwhile. Regardless of the number of meetings, such consultations can be a good opportunity to present any initial energy audit or technical data, to make potential bidders aware of training materials for ESPs that may have been developed, to make them aware of parallel financing programs, to share relevant case studies and any existing M&V protocols, and so forth. This discussion can greatly reduce the potential for misunderstandings later in the bidding process. An agency may also circulate a draft RFP and then hold a meeting to solicit comments and feedback.

Site visits. To help them prepare bids for conducting the IGA and for the performance contract, the ESPs may need to verify some of the technical information provided in the RFP and gather additional, site-specific data to enable them to develop the appropriate technical and cost proposals. Most bidding documents, therefore, contain a provision to allow bidders to conduct site visits. When the procurement covers multiple buildings or facilities,[18] the agency facilitates visits by the bidders to each facility (or a sample of representative facilities). Some bidding documents specify a mandatory site visit (e.g., the Clinton Foundation Johannesburg RFP in South Africa), while in others it is optional (particularly when there are multiple sites).

Oral presentations. Another optional step in the RFP process is the request for oral presentations. Most often, such a step is reserved for the highest-ranked bidders after the evaluation process has largely been completed, as is the case in Japan. It is meant to provide an opportunity for the ESPs to present key features of their proposals and allow agency representatives to ask question about technical and other provisions. The presentation is also often an additional source of points in the overall technical score (see Evaluation section).

Recommendations

On the basis of the case studies and reviews of ESPC experience in many countries, the following are some general recommendations for the development of the RFP:

1. The project must be carefully defined in the RFP. The definition of the "project" in the RFP is one of the most critical steps in the development of that document. Locally applicable public procurement rules and regulations must be consulted early on to determine the level of specificity required, as this move will have implications for the energy audit and other aspects of the process. Where practical, the public agency should try to avoid being too prescriptive in the project specifications, so as to accommodate greater innovation in the ESP responses. Where more parameters are deemed necessary, performance-based specifications, such as target systems or minimum energy savings, are preferred to prescriptive ones, to allow bidding ESPs to propose innovative solutions for the host facility to consider.

2. The process should not be standardized too early. Development of a standard RFP package will facilitate the procurement process and reduce transaction costs for all parties. However, global experience indicates that ESPC public procurement systems will likely evolve based on early experiences. Therefore, standardizing the RFP early may limit the flexibility and natural evolution of the process, as well as of local business models and markets.

3. More steps can be added to the RFP based on local needs. Many countries have created optional steps to help public agencies conduct successful solicitations. These include prequalification or short-listing of ESPs, requiring a detailed audit as part of bid preparation, upstream consultations and pre-bid conferences, site visits, and oral presentations. Such additional steps can make sense in less-developed markets, where additional information can allow both sides to exchange views and information and reduce misunderstandings later in the process. However, these need to be balanced with the management capacity of the client agency, the time planned for the bidding process, and the complexity of the facility.

4. A number of countries have been moving toward prequalifying or even certifying ESPs to minimize the technical aspects of the proposal review and selection process. Prequalifying or short-listing potential bidders can expedite the technical selection process, while also reducing the costs to ESPs of preparing detailed proposals. Short-listing can reduce the level of competition, and prequalification is therefore generally preferred. Also, an overemphasis on the technical qualifications of an ESP bidder should not result in underappreciation of the merits of its technical proposal. There is no consensus on the certification of ESPs. Whereas some certification from a credible agency can greatly raise the level of comfort of potential clients, it can be a barrier to market entry for new ESPs but the certification process require strong technical due diligence and robust governance arrangements.

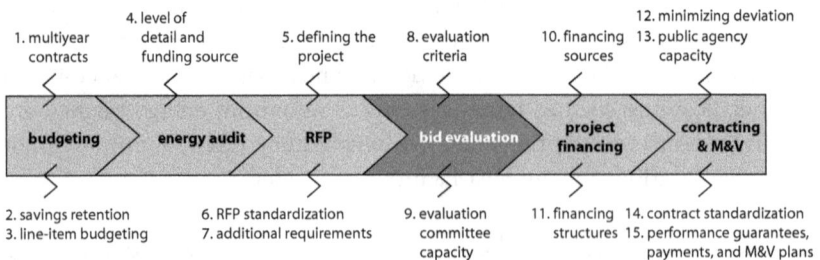

Bid Evaluation

ESP proposals can be very complex to evaluate. First, proposals will generally offer different solutions with varying degrees of energy savings, costs, and risks. Second, ESPC projects have many stages—audit, detailed design, financing, procurement and installation, commissioning, M&V, and maintenance—each one requiring a different set of skills. Third, the financial proposals involve multiple parameters, including the total investment cost, share of savings given to the public agency, proposed duration of the contract, and rate of return of the project. Therefore, unlike typical services or goods procurement, the cost is not represented by a single value; it thus requires a special method of evaluation.

The evaluation of the proposals depends on the other steps in the procurement process. For example, the more prescriptive the project description in the RFP, the less variation in technical proposals there will be, and thus the easier it will be to evaluate them. Similarly, if the draft contract in the RFP has strict parameters regarding the investment size, contract term, share of savings to the client, and so on, the evaluation process will be more straightforward. Conversely, if the client allows for more flexibility in energy savings solutions from the ESP bidders, evaluating the resulting, dissimilar bids in a clear and transparent manner will be more complex.

Designing a fair and transparent process for evaluating ESP proposals in the ESPC process for the public sector is critical to a successful procurement. Issues in the evaluation process are the following:

Issue #8: What evaluation criteria will permit a fair and transparent comparison, particularly given the variability of technical solutions and multiple parameters in the financial proposals?

Issue #9: How can the government ensure that each public agency interested in ESPC procurement has sufficient technical skills to carry out a proper evaluation?

Evaluation Criteria

A clear and transparent scoring method for evaluating the information supplied in bidder proposals is generally required for all RFPs. An important consideration in the selection process is the amount of emphasis to place on the various aspects of these proposals. The evaluation process for ESPs must anticipate that the proposals from the bidders will be quite different in terms of the target systems, the solutions proposed, the energy savings, and the investment cost, and solutions should recognize that there is no textbook method for determining appropriate measures or for estimating savings. A numerical scoring methodology, that involves the weighting of various criteria, conducted independently by a committee of reviewers, helps to overcome the difficulty of evaluating dissimilar bids. Table 6.3

Table 6.3   Evaluation Scoring Sheet for Proposals

Category	Maximum point value	Weighting factor
Financial: payback period; interest rate charges; cost breakdown; buyout option	100 points	25% (0.25)
Technical: completeness of energy savings estimate; baseline; engineering approach	100 points	25% (0.25)
Implementation: plan for purchasing an installing improvements; monitoring savings	100 points	20% (0.20)
Operation and Maintenance: preventive maintenance approach	100 points	10% (0.10)
Project Management: qualifications of personnel; external sources	100 points	10% (0.10)
Training: Approach for delivering training to facility staff; training cost breakdown	100 points	10% (0.10)
Total score	-	-

Source: Natural Resources Canada, Office of Energy Efficiency.

presents a sample scoring sheet with weighting factors from Canada's Federal Buildings Initiative.

In most countries, an evaluation committee comprising agency representatives and possibly designees from other, oversight agencies with previous experience is appointed to review and score each of the proposals. Many countries, including India, have adopted a two-stage ESP selection process for ESPCs in public facilities. In the first stage, the technical proposals are evaluated using predefined criteria, as previously discussed. This technical evaluation is conducted prior to opening the financial proposals. After the technical evaluation of all proposals is completed, the financial bids are opened (second stage), but only for the bidders meeting the minimum technical score. As presented in Box 6.5, the selection of the winning ESP is based on total benefits to the public client over the life of the project. It is important to introduce life cycle costing methodologies into the procurement of equipment, so that future energy savings are factored into the evaluation. All other costs associated with the project, including fuel costs, O&M, and any financing costs also need to be considered in this analysis. Some countries include oral presentations as part of their evaluation process. Japan, for example, invites those ESPs with the highest rankings, based on their written proposals, to make oral

Box 6.5

Two-Stage Bid Evaluation—an Indian Case

Phase 1 – Technical Evaluation

The first stage evaluation of responsive bids shall be carried out based on the following:

1. ESCO Experience
   - *Qualifications and experience of the staff assigned to the project*
   - *Experience with energy performance contracts (list a maximum of six projects along with references)*
   - *Experience and qualifications of the subcontractors (if any)*

Remarks: In case of joint venture, 60 percent weight shall be given to the lead firm and 40 percent given to the joint venture partners.

2. Project Management Approach
   - *Methodology for conducting the investment grade audit (IGA)*
   - *Project implementation plan for installing selected/approved measures*
   - *Work program in bar chart form, showing the activities during the IGA and activities after submission of IGA, until the signing of the energy performance contract*
   - *Access to and availability of key personnel (on-site, local, regional, national)*
   - *Site organization (site personnel, coordination with municipality, site organization chart, etc.)*
   - *Proposed operation and maintenance approach*
   - *Proposed training of municipal staff*
   - *Methodology for transferring the ownership of the equipment and materials to the municipality on completion of the contract period*
   - *Performance monitoring mechanism and reporting schedule as per the IGA guidelines provided in the bid document.*

3. Technical Proposal
   - *Understanding the municipality's conditions and systems*
   - *Approach to measurement and verification*
   - *Proposed risk and responsibility matrix (risk-sharing mechanism)*
   - *Quality of the sample similar audit provided by the ESCO*
   - *Quality of method suggested to establish the energy baseline*
   - *Quality of calculation procedures, credibility of assumptions used in sample audit*

(continued)

Box 6.5 *(continued)*

- *Accuracy, reliability, and appropriateness of methods suggested for measurement and verification of energy savings*
- *Quality of methodology in identifying and configuring the proposed energy saving measure*
- *Specific approaches that will be used to minimize the disturbance to system during the implementation of the energy efficiency improvements*

Bidders scoring more than the minimum qualifying marks of 70 percent at this stage shall be eligible for evaluation of the financial proposals.

Phase 2 – Financial Evaluation

Prerequisite: The ESCO's bid will be evaluated only if it agrees to a guaranteed minimum energy savings of 20 percent of the agreed-upon baseline, based on the preliminary audit report information and past experience.

The financial evaluation of bids is primarily based on the following criteria:

1. The percentage of savings that the ESCO offers to share with the municipality, which should be 20 percent.
2. The minimum upfront investment that the ESCO proposes to bring in for this project from its own resources.
3. Access to, and ability to raise, finances for implementation of the project, based on the ESCO's balance sheet.

Source: GUDC 2008.

presentations to the agency's evaluation committee on their particular approaches to the project. A combination of the proposal score and the oral interview determines the ultimate winning bidder selected for the project. A consensus of the committee is typically needed to select the winning firm.

Evaluating the technical aspects. The goal of the technical selection process should be to acknowledge the most proven and applicable technology, which will be provided by a reputable vendor, installed by qualified technicians, and has minimal O&M requirements, so as to ensure persistence of savings over a long useful life. Bidders often submit varying technical approaches and engineering strategies to achieve the energy saving impacts. These approaches may derive from the particular bidder's comfort level with a technology, past experience with an energy saving measure, or affiliation with a product supplier, as well as the level of risk the bidder is willing to accept.

The submitted proposals will often be required to provide some threshold level of technical information to be accepted and considered for selection. This can be in the form of a clearly presented description of how the ESP would develop and implement an energy efficiency project. The proposal may also request some prior level of technical experience with the energy saving technologies or previous involvement with the type of public facilities in the project. As noted in the Request for Proposals section, RFPs may also require a minimum level of savings for proposals to receive consideration. This level must be clearly defined in the RFP, and a proposal not meeting the requirements will be rejected from consideration.

To establish some minimal level of technical capability and expedite the technical process, a number of countries, such as India and the United States, have created a qualified bidders list of prescreened ESPs capable of carrying out ESPCs. Precertifying or prequalifying bidders can facilitate technical due diligence and limit time wasted on unqualified bidders. However, some countries with less-developed ESP markets, such as the Czech Republic and South Africa, do not rely on prescreened lists of qualified bidders. In these situations, consideration might be given to bundling multiple public facilities together to attract a wider range of ESP types, including multinational ESPs and large equipment vendors. The multinational ESPs may prefer to team with local partners, or can be encouraged by the evaluation criteria to do so, which can then help the development of local ESPs. To encourage competition, some countries offer scoring points for such joint ventures, as well as for local small businesses, minority firms, and women-owned businesses, or countries use other means to recognize the importance of local providers of goods and services.

The World Bank's Management Services RFP recommends applying technical criteria, such as eligibility, financial performance, company experience, and key personnel, at the prequalification stage. For the technical proposals, criteria such as the methodology, work plan, and staffing plan are recommended.

Determining overall best value. Because the financial dimension of ESPCs can be complex, proposed contract value cannot be the sole consideration. Typical ESPC projects have multiple financial parameters, including the total investment cost, project rate of return, total energy and cost savings, client share of savings, equipment life, and contract duration. Therefore, the determination of the overall best value to the client agency should be the prevailing goal. Some countries, such as Canada, the Czech Republic, India, Japan, and the United States (New York State Energy Research and Development Authority), use a weighted average of multiple cost criteria. Others, including Austria and Germany, rely more on a single financial parameter or calculation (e.g., the net present value, or NPV). An advantage of the single value is that proposals that meet the technical criteria or minimum score can then be opened in public, and the NPV read directly from the proposal, as is done with many service and goods proposals to allow for greater transparency in the public procurement process. Still others use direct negotiations (e.g., Croatia, Hungary, and U.S. FEMP utility energy services contracts). Regardless of the approach, the procedure must be specified clearly in the RFP and should promote competition and fairness, fit with local regulations, and meet client agency needs and capabilities.

A number of countries recognize that among multiple financial criteria, some may be more important to an agency than others. Therefore, rather than use a common method, they have developed templates to allow each agency to assign different weighting to each of the financial parameters (see Box 6.6 on the Czech Republic; also see Case Study 5). Flexible weighting has the advantage of allowing agencies to assign weights based on their preferences and objectives. For example, agencies with a strong environmental performance goal may seek to maximize the energy savings, even if it means including some measures with longer payback periods. They may thus choose to assign greater weight to energy savings than to the rate of return. Others may prefer to maximize the economic benefit to the agency, regardless of the amount of energy savings. A drawback with this approach is that some agencies may find weighting financial indicators unnecessarily complicated and may not possess sufficient expertise to know what the optimal weighting should be to determine the best value to the agency.

In Austria and Germany, the financial evaluation in ESP selection is similarly based on the guaranteed energy savings over the life of the contract, guaranteed investment, and the client's share of the savings above the amount guaranteed, as well as on the structure of the offer, project management, CO_2 reduction, and so on. To facilitate evaluation, the monetary criterion is often expressed as an NPV. The net savings are computed as the energy cost savings, plus the operating cost savings (as percentage of investment cost) and value of investment after the end of

Box 6.6

Retrofitting Public Buildings in the Czech Republic

The Pardubice Region, an administrative unit in the Czech Republic, is located mainly in the eastern part of its historical region of Bohemia, with some 452 municipalities and 505,000 inhabitants. In 2005, about 30 buildings were selected for energy efficiency improvements, including 15 schools and 4 hospitals. A specialized agent worked with the administrative unit to develop and issue an RFP. The RFP did not include a draft contract; therefore, each bidder was asked to propose its best package of services and contractual terms to the client.

Evaluation criteria included the following:

• Amount of equipment/material investment and quality	28%
• Guaranteed present value of reduced expenditures	20%
• Guaranteed present value of annual operating cost savings	20%
• Quality and reliability of warranties	18%
• Share of excess savings allocated to public agency	8%
• Bid price	6%

Seven firms participated in the process, four submitted proposals, and the contract was awarded in September 2006. The US$5.4 million investment, which targets heating, piping, control systems, insulation and energy management, is expected to result in 23 percent energy savings (US$720,000 annual savings) for a 12-year contract.

Source: SEVEn 2008.

the contract, minus the ESP's share of energy cost savings and the public client contribution to the investment (if applicable). The NPV is the dominating criterion, with about 75 percent weight, and the other criteria receive about 5 percent weight each. Some of the German procurement agents also provide spreadsheet templates to help their public clients compare financial proposals and assess which one offers the best value.

The NPV expresses the estimated stream of costs and benefits, over a set period in current dollars, by discounting equipment with different lifetimes appropriately, adjusting for inflation, and factoring in such parameters as initial investment, energy saved, contract duration, and life of equipment. Since all measures with rates of return above the discount rate would enhance the overall NPV, this method of evaluation also encourages more comprehensive projects and discourages "cream skimming," in which only the highest-return measures are considered. It may also make sense for the agency to provide the NPV formula to the bidders in the RFP and request that they calculate the NPV within their proposal, so that the financial bids can be opened in public, the NPVs read aloud, and the firm with the highest NPV invited for contract negotiations. In the event that the bidder miscalculates the NPV, the bidder must bear responsibility and honor its bid values. Agencies should consider the savings achieved not only over the term of the contract, but also over the life of the equipment, as not doing so can distort the results, particularly for more capital intensive measures such as chiller systems and motors, which have an operational life and can yield measurable savings well beyond the typical contract term.

For complex projects involving multiple energy saving measures (as opposed to simple ones that can be easily specified), France has been using a *competitive dialogue* procedure in its public-private partnership (PPP) contracts. In this process, a comparative analysis of the overall cost, performance, and risk sharing of the various bidders is conducted, and the contract awarded to the ESP that submitted the most economically advantageous proposal. This informal process of soliciting and discussing proposals from various ESPs can address the challenge of assessing alternative offers that may be quite different from each other and may include some unique and possibly proprietary features. However, it requires a certain sophistication on the part of the public agency.

Some agencies tried not listing the evaluation criteria in the RFPs, except to make reference to the importance of the cost-benefit ratio as a major determining factor. In an energy and water savings project at a municipal water utility in South Africa, the approach was to solicit alternative solutions from the private sector to address the identified problem of water wastage and its associated energy impacts. The proposed scope of the work was not specified in the RFP, as the agency had no idea what the private sector might come up with or how much risk firms would be willing to accept. In ultimately selecting the ESP, cost-effectiveness—or price versus benefit—was the most important criterion used (Alliance to Save Energy 2006).

The "open book" model, in which an ESP charges the public agency a time-based rate for its services and a fixed fee for subcontracting (equipment, financing, subcontractors, etc.), was discussed in chapter 4. This model also has implications for financial evaluation. The reason is that because the rates and mark-ups in such proposals reflect more traditional cost elements, they are more straightforward to evaluate.

Because ESPCs are a relatively new concept in many countries, some government finance officers approach them with uncertainty and caution. These officers worry that ESPCs may entail higher costs than are paid for similar goods and services obtained through conventional procurement practices, perhaps without corresponding benefits. Therefore, in Germany, the agency must also evaluate the ESP bids against its traditional project implementation process, to ensure that it will obtain a fair price and the best value for the agency. The bid review process should follow established procurement procedures that provide auditable documentation of due diligence by the agency in determining the appropriateness of the price for the ESPC services, compared with a self-implementation option by the agency.

Evaluation Committee Capacity

The second issue identified during the evaluation step is the often weak technical capacity of public agencies and evaluation committees. If this issue is not addressed, a risk arises that the evaluation committee may not fully understand and be able to evaluate the relative merits, risks, and overall viability of the various technical proposals received. In one of the project case studies reviewed, the evaluation committee unfortunately did not fully understand the performance contract concept and actually selected an ESP that proposed a fixed payment scheme. Once the proposal was accepted, a contract was signed. When the actual energy savings turned out to be less than expected, contractual disputes arose.

A number of countries, such as Canada and the United States, have identified and made available project facilitators—from oversight agencies, national laboratories, or other experienced public organizations— who can help an agency review and assess proposals. It is also common for agencies to retain independent technical experts to help with the various phases of the ESPC, from initial RFP development through proposal review and selection. This outsourced technical support could supplement in-house agency technical staff in preparing the RFP, gathering technical data on the energy-using systems and baseline energy use information, reviewing the submitted proposals, evaluating the technical feasibility of the IGA conducted by the ESP, overseeing the commissioning of the systems following project implementation, and monitoring the M&V reporting by the ESP. Such technical support could prove to be a very worthwhile investment, for at least the first several projects, until an agency's own staff gains experience in dealing with ESPCs or until a resource network of project facilitators is established. Austria, the Czech Republic, and Germany have created a network of quasi-public and non-governmental procurement agents that assist public agencies throughout the procurement process, including advising evaluation committees on the relative merits of ESP proposals.

Recommendations

As with the other steps, each country's existing procurement process and guidelines will guide the evaluation process to be used. Countries and agencies adopt different procedures and have different preferences, and these must be taken into account. Requesting wholesale changes in public procurement rules is generally not advised, so incremental adjustments are recommended.

The criteria and their respective weighting will drive the selection process. Special attention needs to be paid in establishing these factors, as prospective bidders may attempt to "game" the process to maximize their scores when their proposals may not really be the best deal for the agency. Picking the best overall proposal requires the evaluation of dissimilar bids, energy cost saving impacts, and project investments. The two-stage evaluation process appears to have the most merit, as it allows only those proposals with strong technical merits to proceed to the financial evaluation stage. In the financial evaluation, use of a single indicator, such as the net present value, or NPV, also seems the best choice, since it allows for various financial factors to be aggregated into one clear indicator. Use of the NPV (as opposed to the cost-benefit ratio or rate of return) also helps to prevent cream skimming and can enhance transparency by allowing financial bids to be opened in public and the NPVs to be read aloud. If a weighted average of the financial parameters is preferred, the agency should consider the various elements carefully to ensure that it selects the most advantageous proposal based on its needs.

Institutional capacity must also be duly considered. Where experience is limited and agency capacity is weak, access to advisory services, training, standard RFPs, and other support can help ensure a positive outcome to the ESPC procurement process.

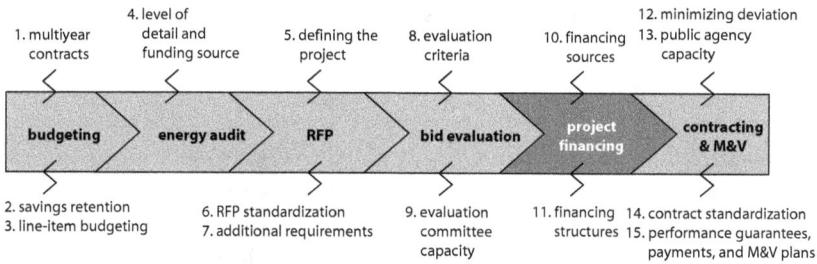

4. level of detail and funding source
1. multiyear contracts
5. defining the project
8. evaluation criteria
10. financing sources
12. minimizing deviation
13. public agency capacity

budgeting → energy audit → RFP → bid evaluation → project financing → contracting & M&V

2. savings retention
3. line-item budgeting
6. RFP standardization
7. additional requirements
9. evaluation committee capacity
11. financing structures
14. contract standardization
15. performance guarantees, payments, and M&V plans

Project Financing

Although financing is not a procurement issue, accessing financing is a critical step in the ESPC process. Although public agencies do expect that ESPs will mobilize financing as part of their bid preparation, in some cases, gaining access to appropriate term financing is difficult. That is why this book includes discussion of how several countries have addressed the financing barrier to public sector ESPCs, for other countries to consider.

During the review, many different mechanisms for financing public sector ESPC projects were identified. Two issues related to project financing emerged that require consideration by public agencies contemplating the use of ESPCs:

Issue #10: What are the possible sources of financing for ESPC projects?
Issue #11: What are the possible financing structures of ESPC projects? Who takes the loan and is responsible for repayment?

Financing Sources

ESPCs must secure project financing to be implemented. Some governments, particularly municipalities, which may be able to access cheaper capital through bonds, may prefer to finance such projects on their own. But other governments are interested in ESPCs specifically for their ability to secure alternative financing for their projects. Mechanisms for financing public sector ESPC projects outside of the public finance option have included the following:

• Requiring the ESP to provide financing through commercial channels
• Creating special energy efficiency funds
• Establishing special public sector agencies to provide or facilitate ESP financing

- Leveraging financing from commercial financial institutions through credit or risk guarantees
- Using carbon financing

Commercial ESP financing. In mature markets such as Canada, Germany, and the United States, ESPs have developed relationships with commercial financial institutions to arrange financing for energy efficiency projects. In these markets, commercial lenders have experience in financing such projects; they understand the nature of the ESP performance guarantees and that the risks of nonperformance rest on the track record of the ESP. If the performance risk is low, the revenue stream to repay the loan, which is generated by the project's savings, is thus considered fairly reliable. Given that public sector credit risk is typically low in these markets, the overall risk profile of the project can be viewed as reasonably low.

In developing countries, however, this model can be difficult to implement. Local ESPs often do not have a track record of delivering successful ESPC projects, local banks may not fully understand the energy efficiency project or the ESPC mechanism, and the public sector agencies may not be as creditworthy. Concerns may also be present, particularly on the part of international ESPs, about the effectiveness of contract laws and whether ESPCs are legally enforceable. Such countries may, therefore, need to consider alternative ways to bring such projects to financial closure. In India, commercial ESP financing has recently been used for projects of the central government, as well as for municipal energy efficiency projects in the state of Tamil Nadu. Because local commercial financial institutions expressed concern about the security and timeliness of payments from the public agencies, a "payment security mechanism" has been established for the projects.[19]

The payment security mechanism involves the establishment of a Trust and Retention Account at a trustee bank by the public agency and the ESP. The public agency makes deposits for energy services into this account, based on baseline consumption patterns. Payments are then made from the account by the trustee to the (utility) energy suppliers and to the ESP, based on the satisfaction of the performance guarantees (see Figure 6.3).

Another mechanism is "ESP receivables financing," also referred to as "factoring" or "forfeiting." It has been used by the European Bank for Reconstruction and Development (EBRD) in Bulgaria and Russia. In these markets, local financial institutions were reluctant to finance individual

Figure 6.3   Use of Payment Security Mechanism in India

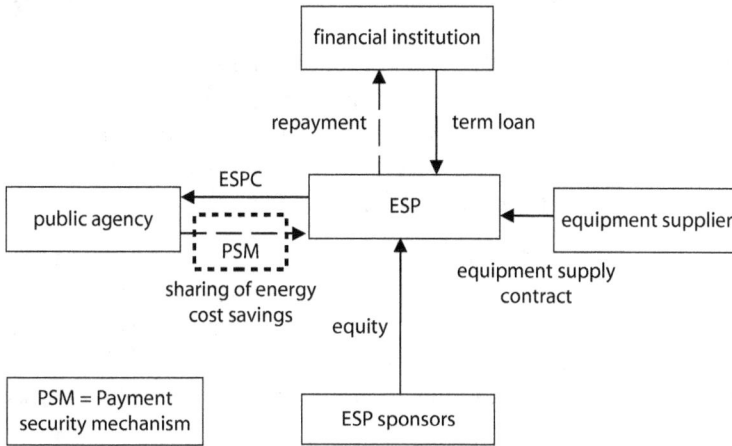

Source: ADB 2004.

projects because the early project development, construction, and commissioning were viewed as risky investments. However, once the ESPC project
was commissioned and performing properly, the risks were substantially
reduced. Therefore, EBRD worked with local financial institutions to
create a commercial financing scheme in which qualified ESPs could sell
their accounts receivable to a local bank, thus being recapitalized to undertake additional ESPC projects. The ESPs were able to access short-term
capital for the early stages of the project and then refinance with longer-
term capital under the EBRD lending programs.

Creation of special energy efficiency funds. Where commercial banks are
unable or unwilling to finance ESPC projects in the public domain, alternative, sustainable financing mechanisms are needed. One approach is the
establishment of special purpose funds dedicated to financing energy efficiency projects. Such funds, often referred to as "energy efficiency funds,"[20]
have been created using a number of different funding sources, for example, a tariff levy on electricity consumption, special taxes, general state tax
revenues, state bonds, petroleum taxes, noncompliance fines from energy
efficiency laws, or certification fees. Where the funds can be operated commercially but require initial capitalization, bonds or general revenues may
make more sense. However, where the fund offers preferential financing
or waives certain audit and other costs, a more reliable and sustainable

source of funding is a tariff levy established by the regulator (i.e., public benefit charges or DSM surcharges) and collected by the utility via the customer's bill.

Such a fund was established in the United States by the State of Utah to finance energy efficiency projects in its public school districts.[21] This fund provides 100 percent interest-free financing (after utility rebates and any other financial incentives) for public sector energy efficiency projects. Other examples include the following:

- The New York State Energy Research and Development Authority (NYSERDA), which has special financing programs for public agencies in the state, including the New York Energy Smart Loan Fund[22]
- The Romanian Energy Efficiency Fund (FREE), which has provided financing with World Bank/GEF support to a number of public sector projects (Voronca 2008)
- The Bulgarian Energy Efficiency Fund (BEEF), which has provided loan financing with World Bank/GEF support to a number of municipal energy efficiency projects (see Box 6.7)
- The Kerala State Energy Conservation Fund (India), being established under the mandate of India's Energy Conservation Act of 2001 to provide financing in the form of grants and loans for public sector energy efficiency projects
- The Korea Energy Management Fund, which has created a low-interest loan fund for ESPC projects (KEMCO 2006)

Although energy efficiency funds have proved useful in stimulating the market, their long-term sustainability needs careful consideration, as well as their ability to work with, and not crowd out, financial institutions that may want to enter the public ESPC financing market. Revolving loan funds, which may be capitalized with public funds but seek to operate on a sustainable basis, may be another option to consider.

Creation of special agencies. Some countries have created publicly owned or PPP special purpose entities, such as the public ESPs discussed previously. Other types of special agencies may be endowed with funds to finance public sector ESPC projects. The South Korean government established the Korea Energy Management Corporation (KEMCO), which in the late 1990s began providing financing for ESPC projects, including ones in the public sector, as well as facilitating the development

Box 6.7

Energy Efficiency Financing in Bulgaria

In Bulgaria, commercial banks were failing to meet the demand for capital from the many financially viable energy efficiency projects because of a lack of liquidity in the capital market and the perceived high risks of energy efficiency investments. Many small and medium-sized enterprises, housing cooperatives, municipalities, hospitals, and other entities thus had very limited access to project financing.

The Bulgarian Energy Efficiency Fund (BEEF) was established in early 2006 as a dedicated, PPP-based finance facility to support energy efficiency improvements in the country. Its initial capitalization was US$15 million, including a US$10 million GEF grant through the World Bank, contributions from the Bulgarian and Austrian governments, and cofinancing from private Bulgarian firms. BEEF is a strictly commercially oriented fund, and it achieved financial self-sufficiency in 2009. BEEF offers three financing products for energy efficiency projects dealing with building rehabilitation, street lighting modernization, small co-generation systems, and other projects:

- Loans to small, bankable energy efficiency projects (up to US$1 million)
- Partial credit guarantees, with up to 80 percent coverage
- Low-cost portfolio guarantees to ESCOs and housing cooperatives, with coverage up to the first 5 percent of delinquent payments in the portfolio

In its first three years of operation, BEEF approved more than 75 energy efficiency projects valued at US$21.9 million, with BEEF financing of US$11.5 million. More than 60 percent of the projects approved are in the public sector. They include municipal buildings, universities, and hospitals, about a quarter of which engaged ESCOs for project development and implementation.

Source: World Bank 2005; Istvan Dobozi, World Bank, pers. comm. 2009.

of the ESP industry. KEMCO is a government agency that receives public funds, which it then provides as loans for ESPC projects, through qualified financial institutions (Lee et al. 2003).

Another interesting example of a special agency is the super-ESCO being established by the government of the Philippines with financing support from the Asian Development Bank (ADB). The EC2 Corporation will act as a traditional ESCO for public sector facilities and directly demonstrate the benefits of energy efficiency projects for schools, hospitals, and other public facilities (ADB 2009). Activities will include (a) preliminary feasibility analysis, including detailed audits of facilities and design of energy efficiency options and financing; (b) installation services and management; and (c) O&M and performance monitoring. EC2 will bear the costs of implementing energy-saving measures, including the costs of energy audits; project design; acquiring, installing, operating, and maintaining equipment; and training operations and maintenance personnel. It will receive a share of the energy savings resulting directly from implementing such measures during the multiyear term of the contract. After EC2 is paid, the remaining savings will be shared between the government facility and the government (the Department of Finance, through reduced budgetary allocations).

Credit or risk guarantees. The International Finance Corporation (IFC), in collaboration with the GEF, has successfully used the strategy of leveraging commercial financing for energy efficiency projects, including ESPCs, using partial credit or risk guarantees. The rationale for this approach, initially implemented under the Hungary Energy Efficiency Co-Financing Program (HEECP) (IFC 2005a; 2005b), and later expanded to six Central and Eastern European countries under the Commercializing Energy Efficiency Finance Program, or CEEF (IFC 2006a; 2006b) was that one of the key reasons local commercial banks were unwilling to finance such projects was their perception that the projects were high-risk. The IFC/GEF approach included providing guarantees as well as technical assistance to the banks to inform and educate them on energy savings project economics and implementation strategies. By 2007, more than 500 guaranteed transactions had been completed in five countries with participation by some 14 banks. USAID has also used its Development Credit Authority credit guarantee mechanism to support municipal energy efficiency projects in Bulgaria (CORE International 2006), resulting in over US$8.1 million of lending for 30 projects, with no defaults. More importantly, the local bank, UBB, now offers loans for up to five years for municipal energy efficiency projects.

The primary objectives of these risk-sharing programs were to expand the availability of commercial financing for energy efficiency projects and to develop a sustainable commercial lending market for energy efficiency investments. The programs developed and used two primary tools to accomplish these objectives: (a) transaction guarantees, or the provision of partial risk guarantees to participating financial institutions for their investments in energy efficiency projects; and (b) technical assistance to strengthen and expand capacity among financial institutions, energy end users, and energy efficiency companies, including ESPs to develop, finance, and implement energy efficiency projects. The success of the programs in Central and Eastern Europe has led to the development of similar programs in China, the Philippines, and Russia.

Carbon financing. One of the benefits of lowering energy consumption is the reduction in greenhouse gas emissions. Projects that contribute to a reduction of such emissions may be eligible to receive carbon credits that can be translated into funding using one of several existing mechanisms. One important way to obtain tradable carbon credits is through the Clean Development Mechanism (CDM), which allows emission reduction projects in developing countries to generate certified emission reductions (CERs) that can be sold and credited to industrialized countries to help them achieve their emissions reduction targets under the Kyoto Protocol.[23]

The basic principles of the CDM are simple. Developed countries can invest in low-cost abatement opportunities in developing countries and receive credit for the resulting emissions reductions, thus lowering the reductions needed within their borders. While the CDM lowers the cost to developed countries of compliance with the protocol, developing countries will benefit as well, not just from the increased investment flows, but also because these investments will advance sustainable development goals.

In general, carbon financing provides only a portion of the funds required for implementation of an energy efficiency project. One example is the municipal street lighting project of the Pune Municipal Corporation (India), which was implemented using an ESPC wherein a portion of the financing is expected to be received from CER revenues. For certain types of projects, such as efficient domestic lighting, the carbon revenues may be sufficient for financing most or all of the project investment.

Public ESPC projects are eligible for carbon financing and, since the ESP must establish a baseline (under the IGA) and implement a measurement and verification plan, some of the additional steps typically associated with carbon financing projects are already included. Carbon financing can thus be an attractive supplemental source of project financing. Further, where specific issues arise in the contracting process, for example, concerning multiyear contracts or perceived public client creditworthiness, carbon finance could be sought to pay the ESP in out-years, allowing the public agency to consider shorter ESP contract durations while providing a secondary (and possibly more stable) source of revenues. It is important to note, however, that while certain CDM transaction costs can be reduced as a result of the baseline and M&V work included in the ESPC approach, several additional costs (relating to validation and registration of the project) and issues may increase financing complexity. These issues make more time-consuming a process that is already potentially difficult for a public entity. They include negotiating agreements for sharing CDM revenues between the public entity and the project developer and establishing an appropriate process for selling the credits realized. A summary of some of the financing schemes can be found in Figure 6.4.

Financing Structures
Two basic structures have been widely used to finance ESPC projects:

- Financing with the public agency as the borrower
- Financing with the ESP as borrower

Under the first approach, the ESP or the public agency arranges financing for the project. The agency enters into an ESPC with the selected ESP and signs a separate financing agreement with a financial institution. The

Figure 6.4   Continuum of Options for ESPC Financing

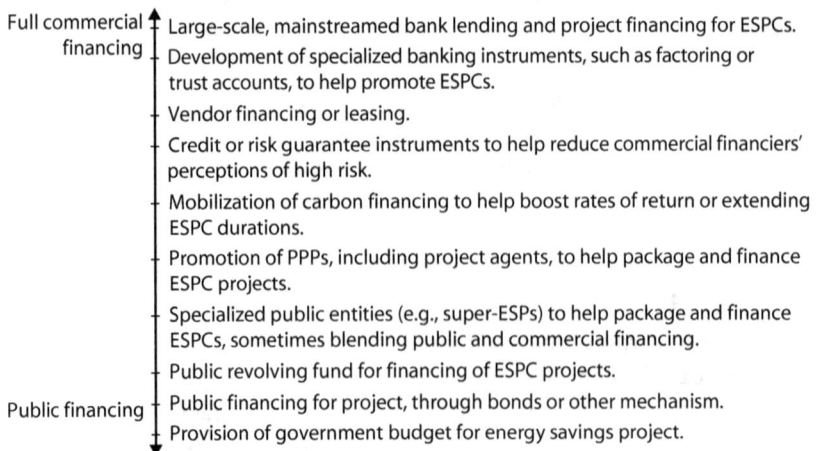

Full commercial financing	Large-scale, mainstreamed bank lending and project financing for ESPCs.
	Development of specialized banking instruments, such as factoring or trust accounts, to help promote ESPCs.
	Vendor financing or leasing.
	Credit or risk guarantee instruments to help reduce commercial financiers' perceptions of high risk.
	Mobilization of carbon financing to help boost rates of return or extending ESPC durations.
	Promotion of PPPs, including project agents, to help package and finance ESPC projects.
	Specialized public entities (e.g., super-ESPs) to help package and finance ESPCs, sometimes blending public and commercial financing.
	Public revolving fund for financing of ESPC projects.
Public financing	Public financing for project, through bonds or other mechanism.
	Provision of government budget for energy savings project.

Source: Author.

ESP is generally not required to put up any equity but is responsible for meeting the performance guarantees agreed upon in the ESPC. The public agency pays off the loan principal and interest from the energy savings. By arranging the financing terms appropriately, the agency can always get a positive cash flow from the energy savings. This approach is commonly used in guaranteed savings ESPC (see Figure 6.5).

In the second approach, the ESP finances 100 percent of the investment through a combination of commercial debt and equity. In developed markets, the ESP typically provides 20 to 30 percent equity[24] and borrows the remainder from a partner financial institution. Often the ESP develops partnerships with commercial lenders, so that as long as the ESP identifies creditworthy clients, or reputable public entities, the lender will approve the loan. The ESP will sign the ESPC with the agency and sign a financing agreement with the financial institution. As with the first approach, the public agency does not have to make any upfront investment and receives a positive cash flow throughout the life of the project, albeit a smaller one than in the previous case since financing payments will be slightly higher. This approach is commonly used in a shared savings ESPC (see Figure 6.6).

These are the main structures for financing ESPC projects, but variations may also be used, such as vendor financing, in which the financial institution provides financing to an equipment supplier, or equipment leasing, with the lease payments based on the expected energy savings. There have also been some recent attempts at aggregating ESPC financing, for example, through special purpose entities, in which several financial institutions and investors provide capital to a vehicle to finance a portfolio of ESPC projects.

Figure 6.5   Commercial Financing with Public Agency as Borrower

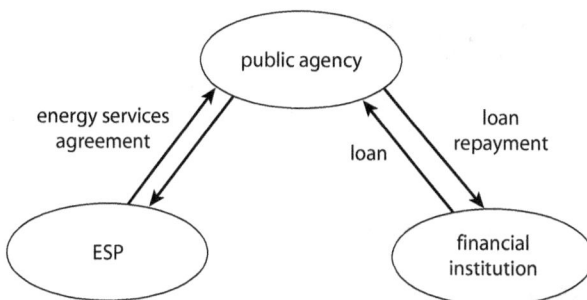

Source: Authors.

Figure 6.6   Commercial Financing with the ESP as Borrower

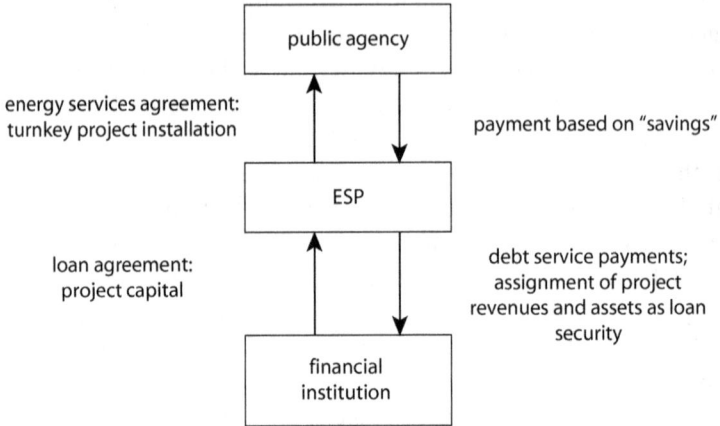

Source: Authors.

Recommendations

The arrangement of financing is a key element of the ESPC process, and a variety of financing sources and mechanisms have been used. Some general recommendations include the following:

1. When a mature commercial financing market exists, with banks actively participating in energy efficiency financing, ESP financing through commercial financial institutions can be the preferred approach. Informational barriers may still exist, however, so initiatives to facilitate access to this financing may be needed.
2. Where commercial financial institutions have limited experience and interest in financing ESPC projects, but liquidity is not an issue, credit or risk guarantee instruments may be more useful in inducing them to participate. Where the commercial financial institution market is immature, an energy efficiency fund or a special purpose agency with funding capacity may be a more suitable approach to stimulating the market and achieving large-scale financing and energy savings.
3. All financing options should contain flexibility on which party will borrow the funds. The programs that have been able to create the most deal flow have typically operated flexibly, trying to develop customized financing options and structures based on each project's needs, ESP and public agency creditworthiness, project cash flows, and other key project parameters.

1. multiyear contracts	4. level of detail and funding source	5. defining the project	8. evaluation criteria	10. financing sources	12. minimizing deviation 13. public agency capacity

budgeting	energy audit	RFP	bid evaluation	project financing	contracting & M&V

2. savings retention 3. line-item budgeting	6. RFP standardization 7. additional requirements	9. evaluation committee capacity	11. financing structures	14. contract standardization 15. performance guarantees, payments, and M&V plans

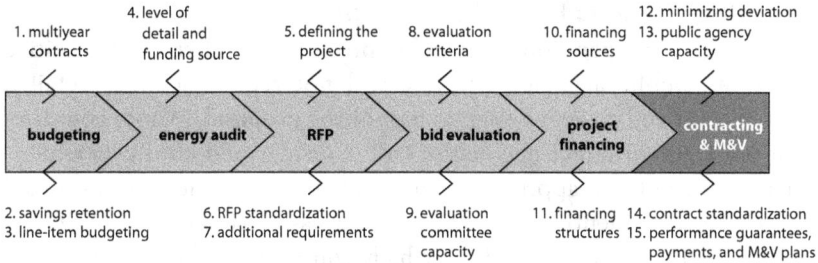

Contracting and Measurement and Verification

The contract between a public agency and an ESP for the provision of energy efficiency services must be structured to address a number of the items discussed previously in the chapter, including performance guarantees, payments, and measurement and verification (M&V) plans. Many of the basic parameters will have to be sorted out early on, since the RFP typically includes a draft contract. However, because some bidders may propose attractive solutions not previously considered by the public agency, adjustments to some of the contract provisions may be needed. Once the highest-ranked proposal is selected, the public agency will, therefore, need to negotiate and finalize ESPC.

The global review identified a number of unique features and challenges in contracting, as outlined below. A particularly unique feature of ESPCs, which makes this step more difficult, is that at the award stage, the exact cost of the project and detailed performance parameters may not yet be determined, as some of the details will be confirmed only after the IGA has been performed. Issues at this stage, therefore, include the following:

Issue #12: How can the public agency ensure that the final ESPC does not deviate too much from the ESP proposal, since the IGA is often performed after the contract has been awarded?

Issue #13: How can the public agency enhance its capacity to support contract negotiations and to supervise both the contract and operations and maintenance (O&M)?

Issue #14: Should the public ESPC be standardized? If so, who should standardize the key documents such as performance contracts, M&V protocols, and the like, and when?

Issue #15: What is the nature of typical performance guarantee clauses and M&V plans?

Minimizing Deviation from the ESP Proposal

To reduce bid preparation costs and improve the number of competitive proposals, public agencies typically will not require a final, detailed energy audit and project design as part of the proposal. (Those countries that use IQCs or direct negotiations, of course, would not likely experience this issue.) This approach means, then, that once the highest-ranked bidder has been identified and the contract has been signed, the ESP will be required to conduct an IGA, which will provide the level of detail necessary to complete the project design, determine the final investment amount, develop an engineering estimate of the energy savings, determine the final ESPC parameters, and establish the baseline from which the project savings will be computed.

Since the full technical performance of each system is not known until the IGA is performed, bidding ESPs provide only a best estimate in their proposals. This approach creates a potential risk for the public agency, as ESPs may have an incentive to "over-promise" at the proposal stage to win the contract, and then may reduce the final energy savings estimate based on the IGA findings. To prevent that from occurring, contracts are usually awarded in two phases. In the first phase, the ESP conducts the IGA and submits it for approval by the host facility. Often the agency will allow only a small variation in guaranteed savings at the IGA stage from the ESP's proposal (under the previous U.S. FEMP IQC, variances could not exceed 20 percent; the new IQC allows agencies to specify this average). Variations greater than the limit may risk the contract's being voided, in which case the ESP would not be reimbursed for the cost of the IGA. The initial ESPCs in Brazil allowed the selected ESPs, once the IGA had been completed, to increase the amount of energy cost savings by up to 20 percent or reduce savings by no more than 10 percent without penalty. In Gujarat, India, the RFP required the winning ESP to honor the energy savings amount in its proposal, regardless of the findings of the IGA. In the event the agency chooses not to proceed with the project after the IGA has been received and approved, the agency generally must reimburse the ESP for the full cost of the IGA. The IGA is also often subject to certain confidentiality clauses, since some of the information may be proprietary. Therefore, agencies need to have a reasonable level of credible data upfront, and the RFP needs to encourage ESPs to provide conservative yet competitive proposals to avoid such situations.

Similarly, under these two-part contracts public agencies need assurances that the ESP cannot inflate project costs after the IGA has been conducted. One option is for the RFP to require the selected ESP to maintain a comparable net present value (from the agency's perspective, with a small deviation permitted) to the original proposal. In this way, if the energy savings are slightly less, then the share of savings to the agency could be slightly increased. Alternatively, the RFP can require unit prices for specific work elements or equipment, such as price per lamp replaced or per control unit installed, as was done in Hungary. Another approach is "open book" contracting, under which the ESP agrees to provide the public agency detailed information on the costs of implementing the energy efficiency measures and to be compensated on the basis of the cost plus a prespecified percentage fee or markup.

Public Agency Capacity

Many public agencies in developing countries have little or no experience with the ESPC approach and may, therefore, find themselves at a disadvantage when negotiating such a contract with an ESP. A need thus arises for training and capacity building in the area of ESPC negotiations and supervision. As discussed in chapter 4, governments have addressed this issue of building capacity for contracting in a number of ways:

- *Preparing ESP IQCs.* The IQC (or indefinite quantity contract) is a "master contract" that is prepared by a government agency and signed with one or more ESPs, as has been done in Canada, Hungary, and the United States. Such contracts define the general terms and conditions of an ESPC without defining the specific nature and quantity of the services to be provided. The agency and the ESP then simply have to define the scope of services and use the standard IQC contract.

- *Creating special purpose companies.* Some governments have established public entities, such as publicly owned ESPs (e.g., super-ESPs, utility ESPs, PICOs), as a way to take a more active role in promoting ESPCs. They then build capacity within these fully or partially owned public ESPs to deal with contracting issues. This capacity has been done in Belgium, Croatia, the Philippines, and Ukraine.

- *Employing a procurement agent.* This approach involves engaging a special procurement agent to facilitate the contracting process, as has been done in Austria, the Czech Republic, Germany, and the Slovak Republic. A procurement agent may be another public agency, a utility, a PPP, an NGO, or a private consulting firm that assists the public agency, typically on a fee-for-service basis, through the entire ESPC procurement process, including negotiations and contract supervision.

- *Bundling public projects together.* Multiple public agency energy efficiency upgrade projects are bundled into a single ESPC procurement, and a parent or umbrella agency addresses the various aspects of the ESPC process, including contract negotiations and supervision (e.g., Tamil Nadu and Gujarat in India; South Africa). Such a process can be more complex, but the incremental cost of adding facilities to the bundle may be low, and the larger transaction may attract new and larger ESPs into the market.

Operations and Maintenance (O&M) training. The maintenance of equipment that the ESP installs to achieve the savings may be performed by the ESP under a maintenance contract or by the facility staff themselves. The specifics of maintenance need to be stated in the RFP and negotiated as a part of the ESPC. The preferred approach varies depending on the circumstances. If the facility has limited technical skills, then the agency may require the ESP either to take over O&M for the contract period or to provide O&M training to the facility staff. Some training will be required in any case, since eventually the equipment will be transferred to the agency. If the ESPC uses a shared savings approach, the ESP and host facility will each have an incentive to maximize the savings and thus may prefer that the contract include maintenance by the ESP.

When the facility staff is to be responsible for operating and maintaining the equipment that the ESP installs, the ESPC needs to include provisions for the ESP to provide training. It is in the ESP's interest to provide proper training, particularly when its payments are linked to the savings. The contract should include an outline of the required training programs or a summary of what facilities managers and operators must do to achieve and maintain the savings. Proper training and instruction are essential to assure that the savings achieved are sustainable during and beyond the term of the contract. Because ESPCs can span many years, provisions may be needed to deal with ongoing training needs, staff turnover, outsourcing, and the like.

It is also important to keep in mind that maintenance issues can be complicated by a number of factors, such as

- Existing maintenance contracts with third parties;
- The risk that the facility's existing maintenance staff may be made redundant (or may need to be retrained and relocated) if the ESP takes over maintenance responsibilities; or

- Lack of incentive for the facility's existing maintenance staff to implement new procedures, undertake training, or take on additional work.

None of the previously stated items is a deal-breaker, and they can generally be addressed in a reasonable manner in the ESPC to the mutual satisfaction of both the public agency and the ESP. However, these issues should be addressed upfront and provisions included in the ESPC, if not in the RFP.

Contract Standardization

To reduce the administrative burden on the agencies interested in using the ESPC approach and to facilitate negotiation and execution of them, many countries have developed standardized ESPCs. Table 6.4 provides some illustrative examples of countries that have developed standard contracts. The U.S. FEMP[25] and the Canadian FBI[26] programs have developed standard ESPCs that are used by federal government agencies in the two countries. Also, the U.S. Energy Services Coalition has developed a model performance contract for state and local government agencies.[27]

In India, the Bureau of Energy Efficiency (BEE) has developed a standard ESPC (with IFC assistance) for municipal agencies (IFC 2008). In Australia, a standard ESPC has been developed for the Energy Efficiency Best Practice Program in the Australian Department of Industry Science

Table 6.4   Standard Performance Contracts

Countries	Approach	Status
Canada-FBI, India, Japan, U.S.	Nodal agencies developed standard contracts for use by government agencies.	Standard contracts are available and have been used in many cases.
Australia	Standard contract was developed by ESP Association.	Standard contract are available and being used.
France	Unique contracting process (PPP) were individually negotiated.	Standard contracts are under development.
Canada-Quebec, Czech Republic, Germany	Assistance and guidance were from NGOs/associations/agents in contracting process.	Standardization is likely to occur with additional experience.
Arab Rep. of Egypt, China, Mexico, South Africa	Little or no effort was devoted to standard contracts.	No standard contracts are currently available.

Source: Compiled by authors from World Bank data.

and Resources by the Australasian Energy Performance Contracting Association (AEPCA 2004). In some cases, such as the Czech Republic, Province of Quebec in Canada, and Germany, assistance and guidance have been provided to public agencies by NGOs, trade associations, or others in developing performance contracts.

In France, the approach used for performance contracting by public agencies is the newly established PPP process. PPP uses a procedure called the *competitive dialogue*, in which the public agency engages in an informal process of soliciting and discussing proposals from various ESPs. The contract is awarded to the candidate having submitted the most economically advantageous proposal. While the PPP process specifies that the contract must include certain mandatory clauses, there is no standard contract. In other countries, such as South Africa, China, Mexico, and Egypt, there have been some ESPCs in the public sector but standard contracts have not yet been developed.

As in the case of the standard RFP, there are substantial differences in the standard contract from one country to another, and contracts have to be customized to the countries' legal systems. The main advantage of standardizing the contracts is the improved efficiency of the contracting process and the lower burden on the public agency staff in contract negotiation. ESPs also prefer more consistency in the bidding process. However, many ESPC projects are unique and may require special contract clauses or terms and conditions. Prescribing a standard contract may impose certain complexities or limitations in the contract negotiations if such special conditions exist. Rather than create standardized ESPCs too early in the development of the market, it may be wise to allow for different types of clauses and provisions to be tested before settling on uniform approaches.

M&V protocols, however, may not be as country specific, since they are based more on the technical aspects of the project than the legal and regulatory issues. Many projects involving international donors have utilized the M&V options specified in the International Performance Measurement and Verification Protocol (IPMVP). These may be adjusted appropriately by mutual negotiation for the specific energy efficiency measures being installed (see next section).

Performance Guarantees, Payments, and M&V Plans

A very important feature of the ESPC is the performance guarantee for the work completed, or the specified link between project performance and ESP payments over the term of the contract. This guarantee helps mitigate the risk to the public agency that the project will not generate sufficient savings to enable it to repay any debt incurred for the project. In most cases, the RFP will include a draft version of the ESPC, which will include the nature of the performance guarantee preferred by the public agency. During contract negotiations the ESP may request, and the agency can agree to, adjustment of the ESPC clauses, as appropriate.

In most cases, this performance guarantee will be in the form of guaranteed annual energy cost savings, which must be larger than the debt service payments, regardless of who takes the loan. Performance guarantees may also address:

- Annual energy savings
- Annual energy and cost savings
- Energy savings plus other cost savings (such as O&M costs)
- Energy supply, in cases where some on-site generation, including co-generation and renewables, are included
- Equipment performance or efficiency level

The specific type of performance guarantee depends on the type of project and the energy efficiency measures being implemented; it usually has to be negotiated between the public agency and the ESP. The ESPC will then specify the guarantees provided, the M&V procedures to ensure that the guarantees are met, and the penalties to the ESP if the guarantees are not met (or bonuses if they are exceeded).

ESP payments. Under an ESPC, payment to the ESP is based on project performance. Therefore, if the project performs according to the plan, then payments are made as stipulated in the contract. But if the project does not perform well, or if, as in some cases, project performance exceeds expectations, then the payments are adjusted accordingly.[28] Some governments have raised concerns over such an approach, as the exact government commitment of an ESPC may not be known upfront, particularly if the project exceeds expectations. However, most governments now have some experiences with performance contracts in other sectors, and there is growing interest in expanding their use to improve contractor performance and obtain the best value for money in public contracting.

When the host facility finances a project, typical contract provisions require that the ESP reimburse the agency for any shortfall in the guaranteed savings amount. When the ESP finances the project and a shortfall occurs, the host facility is entitled to reduce payments to the ESP. In situations where the actual savings exceed the guaranteed savings, some

ESPCs may allow the ESP to receive a share of the excess. Other contracts do not allow the ESP to get any more than the amounts specified at the guaranteed savings level, but do allow the ESP to recoup any payments made due to shortfalls in earlier years if later year savings exceed the guaranteed amounts (see Box 6.8 for example from Germany). Since the ESP is taking the performance risk of guaranteeing the savings and is liable to pay any shortfalls, many governments believe it is reasonable to allow the ESP to gain if the savings consistently exceed the guaranteed amount. Also, if the ESP has incentives to exceed the level of the guarantee, the public agency will also benefit from larger energy savings. It should be noted that when the guarantee is for energy savings (as opposed to cost savings), any energy savings shortfalls need to be converted to

---

Box 6.8

ESPC Performance Guarantees in Germany

German ESPCs typically guarantee the minimum level of energy cost savings, usually as a percentage of total energy costs based on established baselines. In the event that actual energy savings fall below this guaranteed level, the ESP is required to compensate the public client for any differences. In practice, federal facilities are generally obligated to pay the bulk of the ESPC payments, usually around 80 percent or roughly the value of the installed hardware, according to the schedule in the contract, until the following contractual year, when all energy bills are available and the savings can be verified. At that point any government over- or underpayment can be identified.

German budgeting regulations make it complicated to account for contractor reimbursements to public clients, but so far there have been no major cases of ESPC projects underperforming, so this has not been an issue. As an added measure of protection against possible shortfalls, public clients usually require ESPs to furnish a bank letter of guarantee (around 5 percent of the total value of their share of the project's guaranteed savings).

When an ESP refinances a project, that is, where the initial commissioning was successful, and the ESP sells the future client payments (restricted to 70 percent to 80 percent of the agreed payments) to a commercial bank, the performance guarantee to the client remains in effect. In cases of refinancing, the bank letter of guarantee by the ESP often is increased to 10 percent.

Source: Meyer 2009; Toivo Miller, Berlin Energy Agency, pers. comm. 2009.

monetary terms using prespecified tariffs, to avoid changes in payment because of tariff adjustments that are beyond the control of either party. Some options showing various levels of performance guarantees appear in Figure 6.7.

Another issue related to ESP payments is the energy price used to determine the project benefits. Because a project's energy cost savings will be based in part on prevailing energy prices, changes in the cost structure will affect payments under the ESPC and therefore need to be addressed. Under the U.S. Federal Energy Management Program, energy prices are generally stipulated based on an agreed escalation rate. If energy costs rise faster or more slowly, the agency will bear the risk (as it would without an ESPC). Most other countries follow a similar approach, in which base energy costs are stipulated in the ESPC, with some estimated increase each year. This is a reasonable approach, since the ESP is really a service provider and is not in the business to speculate on energy pricing. However, determining how the prices are escalated is important, as it will have a significant impact on ESP payments and thus on the cost of the project to the agency.

Measurement and Verification (M&V). As pointed out previously, M&V planning is a critical step in the ESPC process. The overall objective of an M&V plan is to arrive at a balance between cost and accuracy. Because the agency ultimately pays the cost for M&V, it is preferable to consider carefully the level of accuracy needed in light of the cost of achieving that accuracy. It should be noted that complex M&V procedures often lead to more

Figure 6.7   Sample of Options for ESPC Performance Provisions

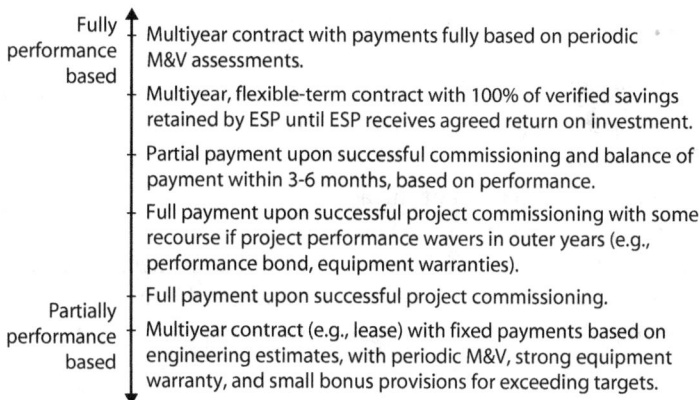

Fully performance based

Multiyear contract with payments fully based on periodic M&V assessments.

Multiyear, flexible-term contract with 100% of verified savings retained by ESP until ESP receives agreed return on investment.

Partial payment upon successful commissioning and balance of payment within 3-6 months, based on performance.

Full payment upon successful project commissioning with some recourse if project performance wavers in outer years (e.g., performance bond, equipment warranties).

Full payment upon successful project commissioning.

Partially performance based

Multiyear contract (e.g., lease) with fixed payments based on engineering estimates, with periodic M&V, strong equipment warranty, and small bonus provisions for exceeding targets.

Source: Authors.

misunderstandings and disputes between the agency and the ESP than simpler ones. Some of the issues related to M&V include the following:

- Who conducts the M&V?
- What is included in the M&V?
- Who pays for the M&V?
- How frequently is M&V conducted?
- How are baseline adjustments to be addressed?

There are three basic options for conducting the M&V:

1. ESP conducts the M&V.
2. The public agency conducts the M&V.
3. The agency or ESP engages an independent third party to conduct the M&V.

Traditionally, public agencies have lacked the capacity to conduct a formal M&V and the ESPs have thus been responsible for it. Examples include Canada and the United States. However, some recent ESPCs in developing countries have required third-party M&V, conducted by highly qualified organizations that specialize in it. The rationale for using a qualified third party is that this option is likely to reduce the possibility of disputes related to satisfaction of the performance guarantees and measurement of actual savings achieved. In South Africa, for example, ESKOM, the national electric utility, has assembled a panel of seven preapproved M&V organizations (all university based), and all ESCOs participating in the ESKOM energy efficiency and DSM program are required to use one of those organizations. In India, recent ESPCs in municipalities have employed third-party M&V agents (usually NGOs or other ESCOs). The IFC *Manual for the Development of Municipal Energy Efficiency Projects* suggests that "an independent expert or service company, not affiliated with any of the contract parties, performs M&V in order to ensure unbiased verification of the achieved savings." (IFC, 2008, p.40).

Regardless of who conducts measurement and verification, it must be done in accordance with the M&V plan specified in the ESPC. Although the RFP may contain a draft plan, the ESP should be required to submit a more detailed one in the proposal and a final plan once the IGA has been completed and the detailed ESPC provisions negotiated. The specific elements of the M&V plan will depend on the nature of the project and individual measures, which is why it is typically finalized only after the completion of the IGA. An illustrative list of key contents of a typical M&V plan is provided in Box 6.9.

A number of different approaches have been used for conducting M&V. Besides the IPMVP and Australasian ESPC M&V guide mentioned previously, other protocols have been developed and used, including the U.S. FEMP *M&V Guidelines* for quantifying the savings resulting from federal projects implemented using ESPCs,[29] the California Public Utilities Commission energy efficiency valuation protocols,[30] and the CDM methodologies approved by the United Nations Framework Convention for Climate Change (UNFCCC).[31]

Most of these protocols provide a range of M&V methods and approaches of varying degrees of complexity. The IPMVP, for example, provides four options (A through D) that may be used for different energy efficiency measures. Some public agencies have used simpler methods with a "deemed savings" approach for M&V, particularly in guaranteed savings contracts. Deemed savings is a predetermined, validated estimate of energy savings attributable to an energy efficiency measure that is used in place of actual M&V activities.[32] Under the deemed savings approach, the public agency and the ESP agree to a

---

Box 6.9

Key Elements of an Measurement and Verification (M&V) Plan

- Description of energy conservation measures, intended results, and the "measurement boundary"
- Documentation of the facility's base year conditions and energy data
- Identification of any planned changes to conditions of the base year
- Identification of the post-retrofit period
- Set of conditions to which all energy measurements will be adjusted
- Specification of M&V options and data analysis
- Procedures, algorithms, and assumptions for performing the statistical validation and anticipated level of accuracy of measurement and results
- Specification of software, budget, and resource requirements
- Documentation and data available for another party to verify reported savings
- Methods for making relevant baseline adjustments for unforeseen changes

Source: AEPCA 2004.

simplified savings calculation procedure and the savings are then "deemed," or calculated, using that procedure. For less-developed markets, use of deemed savings can greatly reduce M&V costs, although it shifts some of the actual project performance risk back to the public client.

It should be noted that if the public agency or the ESPC intends to apply for carbon financing under CDM, the M&V and certification of emission reductions will have to be performed by an independent agency in accordance with one of the UNFCCC-approved CDM methodologies. Several large-scale and small-scale[33] methodologies are available for energy efficiency measures that may be applied under an ESPC project.

An important element in the M&V process is the definition of the baseline, which defines the preimplementation conditions against which the savings (or other specified performance parameters) are calculated. During the term of the ESPC, changes in the defined baseline may occur that affect the level of energy savings or performance. Such changes may include the following:

- Changes in facility use or operating conditions
- Changes in occupancy
- Changes in equipment operating schedules
- Changes in environmental conditions (such as thermostat settings)
- Additions of new energy-using equipment
- Facility refurbishment or rehabilitation

It is important to include in the M&V plan specific provisions regarding what changes would require changing the baseline, how such changes will be identified and tracked, and how the baseline should be modified in case such changes occur. If the project includes carbon financing, any CER revenue shortfalls as a result of project performance must also be assigned in the final ESPC.

The frequency of conducting M&V depends on the project and is negotiated as a part of the M&V plan included in the contract. The frequency may vary from a one-time measurement, to monthly, quarterly, or annual measurements. The ESPC will also specify who will bear the cost for the M&V. When the M&V is conducted by the public agency or a third party engaged by the agency, the cost is generally borne directly by the agency. But, when the ESP conducts the M&V or engages a third party, the cost is often included in the ESP's project costs.

Recommendations

As a public agency gains experience with the ESPC process and RFP quality improves, issues related to contracting will become less important and easier to address. However, when the agency is initially embarking on ESPC procurement, the contracting step can be quite complicated. The following guidelines are recommended:

1. When the ESPC process is new to a country, and many ESPC projects are anticipated, it may be useful to consider establishing a focal agency, a procurement agent, or a public ESP to handle ESPCs in the public sector. Public facilities such as small municipalities or schools, with limited capability and/or resources to enter into their own ESPC, may benefit from bundling of projects under a single ESPC.

2. Development of standardized contract documents will save considerable time and effort in the contracting process. However, it may take some time and experience with ESPCs to develop generally applicable, standard contracts that have demonstrated success and acceptance in a given market.

3. The nature and type of performance guarantees should be defined based on the type of energy efficiency measure being implemented. M&V plans should be explicitly addressed in the contract. Agencies with less experience using ESPCs may benefit from engaging an independent third party to conduct the M&V to minimize the possibility of disputes and protect their interests. The ESPC should also address elements of O&M and training to ensure that the savings are sustainable.

Notes

1. The MTEF process is a multiyear budgeting system that allows government to plan expenditures for a number of years in advance. See http://go.worldbank.org/80OVWNYE30 for more information on MTEF.
2. For details, see DENA (2008). In the states of Brandenburg, Lower Saxony, and Saxony, ESPCs are automatically authorized if profitability can be proved. In Bavaria, Thuringia, and Hessen, ESPCs under certain amounts do not have to be authorized individually.

3. Natural Resources Canada, Office of Energy Efficiency, "Federal Buildings Initiative—Qualification Requirements," 2008.

4. U.S. Department of Energy, Federal Energy Management Program, Master IDIQ contract DE-AM36-09GO290XX.

5. "Works" is defined as construction work (e.g., road/bridge/building construction, pipe laying, installation of transmission towers) and generally does not include goods and services.

6. See http://go.worldbank.org/44Y7O45G93 for the World Bank Management Services RFP template.

7. "DOE Super-ESPC Task Order Request for Proposal (TO RFP) Template," U.S. Department of Energy, November 20, 2008.

8. Natural Resources Canada, FBI Model RFP Documents, http://oee.nrcan.gc .ca/publications/fbi/pdf/2701473E.pdf.

9. The Energy Services Coalition is a national nonprofit organization composed of a network of experts from a wide range of organizations working together at the state and local levels to increase energy efficiency and building upgrades through ESPCs.

10. The Energy Services Coalition, Model Procurement and Contracting Documents, http://www.energyservicescoalition.org/resources/model/index .html#Pre-Qualify.

11. See C40 Cities Climate Leadership Group, http://www.c40cities.org; and city of Johannesburg Metropolitan Municipality, RFP A281, April 3, 2008.

12. See, for example, "Bid Document for Performance Contracting Services to Improve Energy Efficiency at Gujarat Urja Vikas Nigam Ltd. Building," Vadodara, Gujarat, April 2007.

13. A standard RFP document was prepared (in India) under IFC's *Manual for the Development of Municipal Energy Efficiency Projects* and has been used with some modifications in procurements in the Indian states of Tamil Nadu and Gujarat.

14. The World Bank uses the term "prequalification" (as opposed to "short-listing") when the bidding process requires that the applicants meet a minimum set of specific, objective criteria. All applicants who meet these criteria are then invited to bid. Short-listing, as the name implies, restricts the field of bidders to a fixed number (usually four to six).

15. This approach has been used in the procurement of performance contracting services for street lighting and municipal pumping projects by the TNUDF, the Pune Municipal Corporation, and the Gujarat Urban Development Company.

16. The RFP will thus call for the ESCO to "conduct a detailed technical energy audit of the facilities and implement an Energy Performance Contract."

17. Under World Bank procurement guidelines, however, the "pre-bid conference" refers to a meeting with bidders only after the RFP has been issued.

18. For example, the city of Johannesburg RFP included 13 city-owned buildings, and the TNUDF RFP covered eight municipalities.

19. This approach was developed for the public sector in the Asian Development Bank's Energy Efficiency Enhancement Project and has been adapted in a number of public procurements. See ADB 2004. A similar approach was also used in a project at Orissa Sponge Iron Corporation several years earlier.

20. For a review of international energy efficiency funds, see ECO-Asia Clean Development and Climate Program, *Establishment of the Kerala State Energy Conservation Fund, Design Report*, prepared for the Energy Management Centre, Government of Kerala, India, November 2008.

21. *Utah Administrative Code*, Rule No. 638-3, Energy Efficiency Fund, March 1, 2009.

22. New York State Energy Research and Development Authority, http://www.nyserda.org.

23. For more information, see http://wbcarbonfinance.org/Router.cfm?Page=Home&ItemID=24675.

24. In some mature markets, such as the United States, it is not uncommon to finance ESPC projects with 100 percent debt.

25. U.S. Department of Energy, Federal Energy Management Program, Indefinite Delivery/Indefinite Quantity Contract DE-AM36-09GO290XX.

26. Federal Buildings Initiative, FBI Implementation Documents, "Model Energy Management Service Contract," July 20, 1995.

27. Energy Services Coalition, "Energy Performance Contract," http://www.energyservicescoalition.org/resources/model/AttachG-Energy_Performance_Contract.pdf .

28. For example, the USDOE FEMP contract specifies the following: "If the Contractor fails to meet the annual performance requirement as verified by the Measurement and Verification documents, the Government shall adjust the payment schedule to recover the Government's overpayments in the previous year and to reflect the lower performance level into the current year."

29. U.S. Department of Energy, Federal Energy Management Program, *M&V Guidelines: Measurement and Verification for Federal Energy Projects Version 3.0*, April 2008.

30. California Public Utilities Commission, *California Energy Efficiency Evaluation Protocols: Technical, Methodological, and Reporting Requirements for Evaluation Professionals.*

31. See UNFCCC, http://cdm.unfccc.int/methodologies/index.html. For example, methodology AMS II.E is applicable for energy efficiency and fuel switching measures for buildings, AM 0020 for water pumping efficiency improvements, and AMS II.C for energy efficiency activities for specific technologies.

32. NYSERDA and the Texas Public Utilities Commission have used the deemed savings approach for M&V for several of their energy efficiency programs.

33. UNFCCC defines "small-scale" as any project with annual energy savings under 60 Gigawatt-hours.

References

ADB (Asian Development Bank). 2004. *Energy Efficiency Enhancement Project (EEEP), Final Report*. Prepared by Charles River Associates (Asia-Pacific) Pty. Ltd.

———. 2009. "Philippine Energy Efficiency Project. ADB Report and Recommendation of the President to the Board of Directors." ADB Project No. 42001. January.

AEPCA (Australasian Energy Performance Contracting Association). 2004. *A Best Practice Guide to Measurement and Verification of Energy Savings*. Australian Department of Industry, Tourism and Resources. http://www.aepca.asn.au.

Alliance to Save Energy. 2006. "Watergy Case Study: Emfuleni Municipality, South Africa." http://www.watergy.net/resources/casestudies/emfuleni_southafrica.pdf.

CORE International. 2006. "Lessons Learned and Emerging Good Practices in Infrastructure Guarantee Programs under USAID's Development Credit Authority." USAID, April 2006.

GUDC (Gujarat [India] Urban Development Company). 2008. "Implementation of Energy Efficiency Projects in Ten Towns for Municipal Water and Sewerage Pumping Systems through Energy Performance Contract, Request for Proposals." July 30.

IFC (International Finance Corporation). 2005a. "Hungarian Energy Efficiency Co-Financing Program, Project Summary." http://www.ifc.org.

———. 2005b. "Hungarian Energy Efficiency Co-Financing Program, Supplemental Evaluation Report." Prepared by Danish Management Group, May.

———. 2006a. "The Role of IFC in Financing Building Renovations in Hungary." Presentation at DEEM Seminar, Budapest, September 20–21.

———. 2006b. "Commercializing Energy Efficiency Finance, Mid-Term Review." Prepared by Danish Management Group, December.

———. 2008. *Manual for the Development of Municipal Energy Efficiency Projects*.

KEMCO (Korea Energy Management Company). 2006. Annual Report.

Lee, Myung-Kyoon, Hyuna Park, Jongwhan Noh, and J. P. Painuly. 2003. "Promoting Energy Efficiency Financing and ESCOs in Developing Countries: Experiences from Korean ESCO Business." *Journal of Cleaner Production* 11: 651–57.

Meyer, Anke. 2009. "Public Procurement of Energy Efficiency Services: Case Study—Germany." Unpublished World Bank case study.

MWRI (Ministry of Water Resources and Irrigation; Arab Republic of Egypt). 2007. "Requirements and Technical Specifications for the Energy Efficiency Projects at the Ministry" (RFP). Cairo, April.

Natural Resources Canada, Office of Energy Efficiency, "FBI Implementation Documents," Report M27-01-473E, July 1995, http://oee.nrcan.gc.ca/publications/fbi/pdf/2701473E.pdf.

Nexant Inc. 2008. "Lighting Efficiency Improvements Prove Successful in Reducing Operating Costs at the Ministry of Water Resources and Irrigation in Egypt." Sustainable Municipal Energy Services, USAID. Washington, DC: USAID.

SEVEn. 2008."Information on Preparation and Implementation of Energy Efficiency Projects with Investment Means Repaid by Energy Performance Contracting." Unpublished World Bank Czech Republic case study. Washington, DC.

SRC Global Inc. 2005. "A Strategic Framework for Implementation of Energy Efficiency Projects for Indian Water Utilities." Prepared for the World Bank, PPIAF, final report, May. Washington, DC.

Tamil Nadu Urban Infrastructure Financial Services Ltd., "Request for Proposals—Implementation of Municipal Energy efficiency Projects under Performance Contract, Water Supply and Street Lighting," September 2007; GUDC 2008; members of INTESCO Asia Ltd. and Asian Electronics Ltd.

Voronca, M. 2008. "Romanian Energy Efficiency Fund, Present and Perspectives." Presentation at the conference, "Sustainability in Architecture: Dutch-Romanian Ideas on Saving Energy in Buildings." Bucharest, March.

World Bank. 2001. "Thailand ESCO Development Project—Proposal for GEF PDF Block B Grant." Washington, DC: World Bank/GEF, March.

———. 2005. "Bulgaria Energy Efficiency Project Appraisal Document." World Bank Report No. 27545-BUL. February 14. Washington, DC.

CHAPTER 7

Conclusions and Recommendations

The public sector represents a large market for energy services providers (ESPs) and holds significant energy efficiency improvement potential globally. The common ownership and homogeneous nature of many facilities create unique opportunities for large-scale bundling of small projects, which can permit relatively quick implementation and can attract large, new firms into the energy efficiency implementation market. Furthermore, the public sector can often have a catalytic effect on local markets, by demonstrating good behavior to the general public while developing nascent markets for energy efficiency goods and services. The public sector market should thus be actively tapped by policy makers around the world.

To realize the vast energy savings potential is not a simple matter, however. Overcoming restrictive public regulations, poor incentive structures, limited expertise and information, and other obstacles requires concerted effort. Although no simple measures or universally applicable policies have been found, experience from a number of countries shows that delivery of major energy efficiency gains in the public sector is feasible.

As governments pursue a multipronged approach to encourage efficiency improvements in public facilities in all sectors, energy savings performance contracts (ESPCs) are an important tool to be promoted.

ESPCs have a number of advantages for addressing the specific difficulties that public agencies face. The outsourcing of the entire energy efficiency project—from development to financing to monitoring—allows agencies to reap the gains without having to accomplish each step of the project on their own, often entailing multiple procurements taking several years. The ability of ESPCs to offer off-budget financing and to pay for the improvements from the actual energy savings makes the mechanism even more attractive.

Unfortunately, although ESPCs may be well suited to address the challenges of improving public sector energy efficiency, rigid administrative systems are quite poorly suited to simple procurement of energy services. In fact, as this book has shown, each step in the process contains several unique issues that must be dealt with and overcome to reap the benefits of ESPCs. Furthermore, their complex nature requires significant capacity building throughout the public sector.

Helping public agencies manage the ESPC process is a difficult task. The multidisciplinary nature of the issues in each step of the procurement process—from budget regulations, to energy auditing, to public contracting, to project financing—makes navigating those steps challenging. And unfortunately, many procurement and budgeting practices vary from country to country, so solutions must be country specific, as well. Nevertheless, a few basic guidelines are suggested next to help governments determine how to design a suitable procurement scheme for energy efficiency services in their public facilities.

Designing the Right Procurement Process

As illustrated throughout the book, solutions have been developed for all of the issues identified in the public ESPC procurement process. But the barriers each country faces have different features, and what works in one country may not be the right solution for another. It is hoped that among the experiences and solutions described here are ones that can be used as a basis for solving specific situations in other countries or markets.

The case studies revealed a number of different approaches in how governments have dealt with the promotion and procurement of ESPCs in the public sector; from the federal level first (e.g., Canada, United States), to more localized approaches (e.g., Germany), to more donor-driven solutions (e.g., Czech Republic, India). Emerging models have included the IQC umbrella contract mechanism, public ESPs, energy supply contracting, procurement agents, project bundling, and nodal agencies. Each model has its relative merits and drawbacks, so governments need to weigh these and other options carefully.

Each of the steps in the procurement process contains a unique set of issues that can make the process complex. But as noted in chapters 5 and 6, experiences from around the world offer a range of approaches and solutions for countries to consider and adopt or adapt as appropriate. Table 7.1 includes a summary of recommendations for each step identified in this book. By looking at the range of options to address each issue, a country may be able to find an appropriate and feasible solution from another country. Or a country could mix and match, combine, or develop new solutions based on the many approaches used

Table 7.1 Recommendations for Main Public ESPC Procurement Steps

Main steps	Recommendations
Budgeting	Start public ESPC procurement schemes with more autonomous public entities first
	Gain support from, and work with, parent budgeting agencies
	After implementation of a few ESPCs, develop public financing programs to help address budgeting, incentive, and financing issues
	Change the budgeting laws and regulations in the longer term, as required
Energy audit	Consider the level of technical information prospective bidders require to properly define the project
	Where appropriate, provide basic technical data (facility description, equipment inventory, energy bills, etc.) in lieu of energy audit
Bidding documents	Carefully define the project to ensure that it meets local procurement rules and regulations
	Provide broader parameters, such as minimum energy savings or target systems, to avoid being too prescriptive
	Delay the standardization of procurement documents to avoid advance limits on flexibility and the natural evolution of the market (Once sufficient projects have been implemented, standardization can facilitate scale-up and reduce transaction costs.)
	Introduce steps to the bidding process, such as prequalification, detailed audits, pre-bid conference, or oral presentations, based on local needs and capabilities
Evaluation process	Adopt a two-stage evaluation process in which technical proposals are scored first and the highest-ranked proposals proceed to the financial evaluation stage
	Use a net present value (NPV)—or an equivalent single, comprehensive indicator—in the financial evaluation to allow for simple, transparent assessments and limit "cream-skimming"

(continued)

Table 7.1   Recommendations for Main Public ESPC Procurement Steps *(continued)*

Main steps	Recommendations
Financing	In mature capital markets, make efforts to attract commercial financing for ESPCs with informational and other technical support
	Where perceived risks are high, offer credit or risk guarantees to encourage commercial financing of ESPCs
	In immature markets, particularly where liquidity is an issue, create a dedicated energy efficiency fund or other entity to support ESPCs
	Maintain flexibility in all financing programs to allow for maximum market development
Contracting	Designate entities such as nodal agencies, agents, or public ESPs to facilitate public ESPC projects
	Develop standardized contracts that are based on initial ESPCs and that have performed successfully to further facilitate public energy savings projects
	Define performance guarantees based on the type of energy measure being implemented; including an M&V plan in the contract
	Address operations and maintenance (O&M) and client training in the ESPC to ensure that savings persist

Source: Authors.

elsewhere. For example, a municipality may have a strict, one-year contracting limit for public agencies, no funds for an elaborate audit (but the local utility may offer free walk-through audits), flexibility with the RFP process (it need not prescriptively define the project), and access to three- to five-year financing from a public bank. Using the options outlined in Table 7.2, such a municipality may be interested in arranging a walk-through audit by the utility and issuing a flexible RFP, without financing required under the ESPC. It may obtain a loan to finance its project from the public bank and enter into a one-year contract with the winning ESP. Even though the ESP will be finished with the contract after a year, the savings will generally still be sufficient for the municipality to service its debt, and thus it can maintain a positive cash flow. Other measures could be developed to provide additional assurances to the city that the project will perform beyond the one-year term, such as more robust equipment warranties or performance bonds that extend beyond the contract period.

Of course, in reality the process is never quite so simple. The various steps are inextricably linked to one another, making finding solutions more complex. The interlinking steps may mean that a solution to one issue may limit the possible solutions to another. Designing the right process should thus be viewed as a holistic exercise. If one possible solution results in no

Table 7.2 Selecting Solutions from International Experiences

Budget	Audit	Financing	Model	Contract
Progressive	**Prescriptive**	**Commercial**	**High ESP risk**	**Performance based**
Agency's full retention of EE benefits after reform	Detailed energy audit and resulting predefined project	Bank lending and project financing to ESPCs	Full service—shared savings	Multiyear contract and periodic payments based on M&V assessment
Certain autonomy or fixed budget provisions of agency	Mandate audit	Vendor financing or leasing	Energy supply contracting—chauffage, outsourcing, contract energy management	Multiyear, flexible term contract until ESP's agreed return met
Noncash refund to agency from ESPs with retention of EE benefits	Detailed audit from similar, representative facility	Credit or risk guarantee	ESPs with third-party financing—guaranteed savings	Partial payment upon commissioning and balance paid 3–6 months
Partial EE benefits assigned to agency by Ministry of Finance (MOF)	Walk-through audits/evaluation	Carbon financing to boost IRR or extend ESPC duration	ESPs with variable-term contract—first out contract	Multiyear contract and fixed payments with periodic M&V, equipment warranty, and bonus provisions
No agency retention, MOF upfront subsidy/grant/special financing	Institution-led low- or no-cost audit	Financing and packaging by Public-private partnership (PPPs)	Supplier credit	Full payment upon commissioning with some recourse for outer years
No retention but other incentives (e.g., awards, competitions)	Completed audit template	Financing and packaging by public entities (e.g., super-ESPs)	Equipment leasing	Full payment upon commissioning
No retention; MOF mandate on agency EE implementation	Equipment inventory/bill summary	Public revolving fund	Consultant with performance-based payments	Traditional
No retention; ESP procurement by MOF/parent agency	Audit by preselected ESPs under Indefinite quantity contract (IQC) approach	Public financing through public bonds, etc.	Consultant with fixed payments	
	No upfront audit; detailed audit by bidders prior to bid submission	Government budget for EE projects	Low ESP risk	
Restrictive	**Flexible**	**Public**		

Source: Authors.

135

viable solutions for another step, then a new solution to the first issue must be developed. For example, if procurement regulations require a defined project, then the agency may be obligated to have a more detailed, upfront energy audit conducted. Consultation with a wide range of stakeholders is equally important. If possible solutions to an issue are not compatible with potential service providers, financiers, parent public institutions, budget and procurement officers, or technical staff, then it is unlikely that the process will work. Therefore, developing a suitable procurement scheme needs to start with identifying the dominant hurdles first, finding feasible solutions to them, and validating them with other stakeholders. And the process should be sufficiently robust enough that one solution does not undermine another.

Also important is that some difficulties may not be known until the process has been implemented in its entirety. Therefore, rather than seeking to standardize the process too early, or attempting to achieve sweeping changes in existing laws and regulations, simpler, incremental solutions should be developed and tested first, to determine if they actually solve the target issues and to identify other challenges that may not have been seen at the outset. India has embarked on a very useful trajectory for its public procurement programs by testing a few procurements at both the central and state levels, gradually refining the process and enlarging the bundle of facilities each time. By doing that, it allows early issues to be addressed and allows each RFP and contract to be an improved design compared to the previous one. This iterative process of testing and refining allows for a more robust approach to be developed, which will be much better for the country in the long term.

In summary, the right procurement scheme must be developed based on prevailing country, sector, regulatory, and market conditions. Each country need not have a single solution; in fact, many countries, including Canada and the United States, have both federal and state or provincial programs, with quite different procedures and mechanisms. Solutions can be based on the experience of others but generally have to be customized to fit local regulations, markets, and other conditions. And no solution is viable without strong interest and bids from commercial ESPs; therefore, the private sector must be involved in upstream consultations, as well. Some processes will require multiple innovations and iterations before they work smoothly, so governments must persevere and not be discouraged by early setbacks. Significant benefits can accrue if the process is implemented correctly, but governments should not expect the process design to be quick or simple.

Recommendations

As much of the recent literature on the subject has confirmed, energy efficiency is a powerful resource to be exploited but a very difficult one to tap. Developing sustainable markets requires significant government involvement, supportive policies, public incentive and financing schemes, and strong information and outreach initiatives. Perhaps more important, it requires perseverance and patience.

Before officials plunge into the details of the procurement process, it may often be necessary to determine the most viable business and contractual models for ESPs to operate in a particular country. As described in chapter 3, a number of business models have already been developed, and these can provide a useful reference to see which aspects may make sense in a given market. Often it is advisable to begin with simpler models and introduce more complex transactions only as the market develops and supporting systems evolve. ESPC and ESP models developed in the Western countries can be important in understanding the range of options, but they may need to be adapted, and many incremental changes made, before they are workable in many developing countries. Where local ESPs already exist, efforts may be needed to build on successful transactions and to institutionalize the aspects that have worked well. Bundling projects to reduce transaction costs and make the projects more attractive to larger companies, including international ESPs, may also be an approach worth considering.

Working through the procurement process becomes more straightforward once the most viable ESPC models in the local market are known, permitting a clearer idea of the level of sophistication, the ability of ESPs to access financing, their willingness to take on project risks, and so on. Designing the right process will take time, but it is important to get it right. As the market develops, ESPs and the models developed and promoted in the public sector are likely to have significant ripple effects in the private sector, as private firms observe what the public sector is doing and participating ESPs begin to market the services and skills they have refined in public projects.

The following are some key steps for consideration:

1. *Conduct an upfront market survey of ESPs.* Such a survey should be cast broadly to include all types of entities that could serve as ESPs. It should gauge their level of interest in serving the large public sector market and capacity to do so; assess their willingness and ability to take on ESCO-type project risks; evaluate their ability to access

financing; and ascertain the types of contractual provisions they would be most comfortable accepting. The results will feed into the procurement design process.

2. *Hold stakeholder consultations to analyze barriers.* Focus groups or similar events should be set up with various stakeholders involved in typical public procurement processes (e.g., technical and environmental departments, procurement and budget staff, parent budgeting agencies, staff from different types of public agencies, ESPs, financiers) to assess the types of constraints to be expected in the public procurement of ESPCs. These consultations should work through the entire process, from the audit to the end of the contract period, to determine where pitfalls may lie and clearly define their nature. It is also helpful to rank the barriers in terms of those most critical to getting through the process.

3. *Create lists of options to address each barrier.* Potentially viable solutions should be identified for each of the key barriers that the consultations identify, using international experience as a guide. Broad consultations with various agencies can determine if precedents already exist in the market to deal with some of the more difficult challenges. (For example, other agencies may have used multiyear contracts or contracts with performance-based clauses in other sectors, and those could be used to justify some of the proposed ESPC elements.) Possible road maps to completing the procurement process, addressing each of the identified barriers, should be developed with the stakeholders. It is important that solutions to one problem do not undermine solutions to another, and that potential solutions are acceptable to all the relevant parties. At least in the beginning, simple solutions that require small, incremental changes to existing systems are preferred over wholesale changes to public policies or regulations.

4. *Develope and test small procurements.* One or more smaller procurements should be tested to learn whether the process can work in the given market. Some of the solutions may not work in some cases. In other cases, issues not previously identified may present themselves. These difficulties must be worked through if the project is to be implemented. All the challenges should be documented, and all parties enlisted to work collectively to refine the procurement process.

5. *Expand and replicate.* Once the process appears to be working, the scope of the projects can be scaled up. Officials should consider a broader range of systems to be retrofitted, bundle many similar facilities together, consider contracts with longer durations and O&M provisions, and the like, to further improve energy savings results. Processes should be replicated at all levels of government in all sectors.

6. *Institutionalize systems.* In parallel to replication, mechanisms should be created to facilitate the long-term development of the market and process. The design of model templates and documents, the development of longer-term changes to public procurement and budgeting systems to accommodate ESPCs, the creation of nodal agencies and project agents, the launching of incentive schemes and financing programs, the development of public agency targets, the creation of M&V protocols, and so on can help ensure (a) that the process is continually encouraged, (b) that energy efficiency gains are maintained, and (c) that measures are put in place to allow greater potential energy savings to be realized.

PART II

Select Country Case Studies

United States Federal Energy Management Program

The U.S. federal government is the largest energy user in the world, spending $4.3 billion annually for energy in federal buildings and facilities. Executive Order no. 13423, of 2007, and the Energy Independence and Security Act passed by the U.S. Congress in 2007 set new, mandatory federal energy goals. Federal agencies are directed to cut their energy use (compared to 2003) by 3 percent per year through 2015; to significantly increase their use of renewable energy; and to reduce water use by 2 percent per year through 2015. In 2007 it was reported to Congress that the agencies had met the goal of 3 percent reduction in energy use for that year.

The U.S. Department of Energy (DOE) established the Federal Energy Management Program (FEMP), following authorizing legislation in 1995, to help agencies reduce their energy use through energy efficiency and renewable energy installations. FEMP is administered within the Office of Energy Efficiency and Renewable Energy at the Department of Energy. It provides technical assistance, guidance documents, workshops, contracting arrangements, and project facilitation services to federal agencies.

Energy savings performance contracts, or ESPCs, are a major vehicle for agencies to implement energy projects and met the established goals. To a lesser extent, they are also using sole source contractual partnerships with local utilities in utility energy saving contracts, or UESCs.[1] ESPCs allow the federal agencies to implement energy projects and meet the established goals. The original ESPC legislation was enacted in 1985 and gave federal agencies the authority to enter into shared energy savings contracts with private sector ESCOs.[2] Legislation passed by Congress in 1992 further authorized federal agencies to execute multiyear guaranteed savings contracts and directed DOE to promulgate regulations for ESPCs.[3] It also provided the contract authority for UESCs and agency guidance for entering into sole source contracts with local utilities for public utility services, including energy and water conservation. Following a formal rulemaking process, DOE promulgated clarifying regulations for the use of ESPCs in 1995. The procurement process was streamlined in 1998 with the addition of super-ESPCs. A 1999 executive order[4] further strengthened the government's efforts to pursue energy and cost savings. Congress provided permanent authorization for ESPCs in the 2007 Energy Independence and Security Act already mentioned.

Under this array of legislation and regulations regarding ESPCs and UESCs, DOE's Office of the Counsel and other agencies have provided guidance and legal opinions on implementation. For example, agencies can retain 50 percent of energy and water cost savings from appropriated funds for additional energy projects, including employee incentive programs. With the continued pressure to limit federal agency budgets, however, any cost savings that remain after payment of the ESPCs are often cut from the agency's operating budget.

Over the past decade, approximately $2.3 billion in private sector funding has been invested in federal facilities through ESPCs, saving over 18 trillion Btu annually. More than 460 federal ESPC projects had been awarded through fiscal 2007 by 19 different federal agencies in 47 states. The majority of federal ESPC projects have been undertaken by the Department of Defense (48 percent) and the Department of Energy (40 percent) as shown in Table CS 1.1. These projects will save the government $7.1 billion in energy costs, $5.7 billion of which will go to finance the project investment and upgrade facilities, resulting in a net savings to the government of $1.4 billion.

The energy audit. Since about 1998, traditional ESPC projects have entailed agencies' issuing a request for proposals (RFP) from energy service companies, or ESCOs, in a competitive procurement process. Interested ESCOs would conduct an energy audit, at their own expense, and identify improvements that would save energy in the facility.

Table CS 1.1   Agency Participation in ESPC Projects

Departmental agency	Defense	Energy	Justice	Health and human services	Veterans affairs	Homeland security	Other agencies
Percentage of ESPCs (based on $ value)	48	40	6	2	2	1	1

Source: U.S. Department of Energy 2008.

The more recent super-ESPC approach, with prequalified ESCOs, has been designed to accelerate the contracting process. In the new process, an ESCO is competitively selected to service a region or provide a specific type of technology. ESCOs are required to contact the DOE contracting officer prior to submitting initial proposals to agencies. The agency preselects the ESCO under a single source delivery order and issues a notice of intent to award a contract to it.

Types of energy savings projects. ESCO projects under FEMP have covered a broad range of energy saving technologies and include special contracting provisions for certain advanced technologies, such as on-site renewables. The projects often include a mix of quick-payback projects, such as lighting measures, and other measures with longer-term payback periods, such as chillers and motors.

ESPC projects that an ESCO identifies and proposes to an agency unsolicited, outside of a formal bidding cycle, can be awarded as single source contracts, without the usual requirements for competition. However, projects that the agency defines are subject to the same requirements for competition as other government procurements. In these agency-identified projects, the agency defines the technical specifications of a project and assembles a site data package and RFP. Those are issued to multiple ESCOs. The agency then evaluates the multiple ESCO proposals and makes an award.

The more recent super-ESPCs use an indefinite quantity contract (IQC) method, referred to as "indefinite-delivery, indefinite-quantity" (IDIQ) contracts. DOE established this process to make ESPCs a more practical and streamlined tool for agencies to use. These umbrella contracts were competitively awarded to ESCOs that demonstrated their capabilities to provide energy savings projects for federal facilities. The general terms and conditions, along with project requirements and construction standards, are established in the IDIQ contracts, and the agencies implement energy projects by awarding "delivery orders" or "task orders" under the prime super-ESPC. The ESCO can undertake the detailed energy audit only after the agency and ESCO have executed a delivery order.[5] The U.S. Army has a similar umbrella ESCO contracting process established solely for its facilities.

An important component of the super-ESPC procurement process is a justification for waiver of competition, which DOE authorized under legislation (exemption from competition by Federal Acquisition Streamlining Act) that was enacted to encourage ESCOs to initiate projects with agencies. This waiver has been written into the super-ESPC documents and allows contractor-identified proposals. The justification of waiver of competition permits an agency to pursue a single source process in awarding an ESPC delivery order, first conditionally selecting one of the prequalified, multiple-award ESCOs and then accepting an initial proposal from the ESCO for a "contractor-identified" project. Box CS 1.1 summarizes the initial project development phase of selecting the ESCO and developing the delivery order RFP.

Evaluation criteria for selecting the ESCO. Competing ESCOs are evaluated and selected based on their demonstrated capabilities to manage the development and implementation of multiple ESPC projects over a large geographic area and on the technical approach and price of a defined, site-specific project. Two types of ESCOs are participating in the Federal Energy Management Program: DOE-qualified ESCOs and super-ESPC ESCOs. The DOE-qualified list comprises private industry energy service firms that have submitted an application and been qualified by the Qualification Review Board established by the department. The board consists of representatives from the Federal Interagency Energy Management Task Force and DOE FEMP staff. The industry certification process of the National Association of Energy Service Companies (NAESCO) is separate and distinct from the federal requirements.

The Qualification Review Board selects firms for inclusion on the qualified list based on the requirements shown in Table CS 1.2. Almost 100 firms on the DOE Qualified ESCO list are eligible to do the traditional ESPC projects, yet only about a dozen are awarded the super-ESPC contracts. It is not uncommon for firms on the DOE Qualified ESCO list to work as subcontractors of one of the prime ESCOs awarded a super-ESPC contract.

Box CS 1.1

Awarding a Delivery Order to an Energy Service Company (ESCO)

- DOE contracting officer reviews super-ESPC process with agency and opportunity for a project.
- Informal communication exists with ESCOs, reviewing ESCOs' capabilities and experience.
- An ESCO is given approval to submit the initial proposal.
- Agency receives ESCO's initial proposal, evaluates, and determines whether to pursue the project.
- Agency works with ESCO to finalize the delivery order RFP.
- Agency obtains DOE contracting officer review and notice of intent to award.
- Agency issues notice of intent to award to ESCO and issues the delivery order RFP to the ESCO.
- ESCO conducts detailed energy survey and solicits competitive financing offers.
- Agency evaluates ESCO's final proposal of audit findings and financing arrangements.
- Negotiation begins between the agency and ESCO, proceeding until agreement is reached on all issues.
- DOE reviews delivery order and confirms that ESCO meets preaward requirements.
- Agency awards delivery order to ESCO to begin project installation.

Source: U.S. DOE 2007.

Table CS 1.2   Requirements for Becoming a United States Department of Energy (DOE)-Qualified Energy Service Company (ESCO)

ESCO requirement	Description
Prior ESPC or energy project experience	ESCO has provided services to at least 2 clients and possesses experience to successfully implement technologies.
Satisfactory client ratings	Previous project clients provide ratings of "fair" or better.
Financial stability	Firm or its principals have not been insolvent or bankrupt within the past 5 years.
Eligibility for procurement programs	Firm or its principals have not been debarred by the federal government.
In good standing	No adverse information exists that firm is not qualified to perform the ESPC.

Source: U.S. Department of Energy.

The super-ESPCs were awarded to the prime contractor ESCOs in compliance with the Federal Acquisition Regulation procurement requirements for competition. With these contracts in place, the majority of the procurement process is already done, and agencies move directly to developing an energy project. This streamlined process for gaining access to the expertise and private financing that ESCOs offer can save agencies time and money. A super-ESPC delivery order can be awarded in 4 to 12 months, while a typical stand-alone ESPC can take 2 to 3 years (or longer).

On December 18, 2008, DOE announced that it will award 16 new IDIQ ESPCs that will replace the old super-ESPC contracts. The new contracts provide for a maximum individual contract value of up to $5 billion over the life of the contract, eliminate technology-specific restrictions, and allow federal agencies to use the contracts in national and international federal buildings. The agency contract officer must provide each IDIQ contractor a fair opportunity to be considered for any project-specific task order award. A federal agency may select an IDIQ contractor that represents the best value to the agency, using the method summarized in Table CS 1.3.

Multiyear energy performance contracts. The authorization for federal agencies to enter into ESPCs specifies that savings guarantees are mandatory and that measurement and verification (M&V) protocols will be used to verify that the guaranteed savings are realized. The ESCO and the

Table CS 1.3   Fair Opportunity Consideration Award Methods

Contractor-initiated method	An IDIQ contractor proposes a task order for a specific project. If it is acceptable, an agency contract officer issues a written notification of approval.
	Contractor submits proposal with a preliminary assessment (~20 pages) that includes project merits, technical feasibility, level of projected savings, economics, and project cost.
	Agency contract officer publishes notice that the agency has received a task order proposal and invites competing proposals from other IDIQ contractors.
	If no responses, agency may award a task order to the initial IDIQ proposer; contract officer may require investment grade audit to confirm technical and price proposals, before awarding task order. If responses are received, the agency contract officer determines whether to issue award to initial contractor or pursue method presented below to select a different contractor.

(continued)

Table CS 1.3   Fair Opportunity Consideration Award Methods *(continued)*

Government-initiated method	Each IDIQ contractor receives notice of agency's needs.
	Agency contract officer has broad discretion in developing solicitation procedures.
	Formal evaluation plans and scoring are not required.
	Initial contractor selection is based on the preliminary assessment and conditioned on investment grade audit confirming the estimates it contains.
	Agency is not responsible for any costs of preliminary assessment or investment grade audit, unless task order is awarded.
Exceptions to fair opportunity (contract officer has discretion if one of these exemptions applies)	Agency's needs are urgent and providing fair opportunity would result in unacceptable delays.
	Only one contractor is capable of providing the needed services.
	The new work is a logical follow-on to an order already issued.

Source: U.S. Department of Energy.

customer agree on annual, firm-fixed-price payments that are less than the guaranteed annual energy cost savings. The savings to the agency must exceed payments to the ESCO in every year of the contract's term. The contract term for federal ESPCs, including the construction period, may be a maximum of 25 years; the average to date is approximately 15 years. They can include a broad portfolio of energy savings measures, ranging from quick-impact measures to ones with longer payback periods, even up to 30 or 40 years. With the aggressive directives to reduce energy use, agencies have usually implemented many low-cost and no-cost measures and look to the ESPC projects to replace old, inefficient equipment. Projects thus tend to be capital intensive and do not involve "cream skimming," or only quick-payback measures.

Project financing. There are two ways to structure super-ESPC financing—separate construction financing and permanent financing, and escrow financing (Table CS 1.4). One of the principal differences between the two is that escrow financing allows the permanent financing to be placed on the award date, before construction begins, rather than at project acceptance. Depending on the specific financing terms, escrow financing has been used more often by the agencies. In either case, the financing is based on the ESCO's balance sheet.

ESCOs commonly obtain financing from a third party. They must solicit competitive financing offers for the energy project in commercial

Table CS 1.4   Super-ESPC Financing Approaches

Financing method	Description
Two separate loans, one for construction financing and another for permanent financing	ESCO takes short-term loan to fund construction.
	Interest is paid only on funds already drawn as construction milestones are met.
	Construction loan is paid off by permanent financing at project acceptance.
	The two loans generally involve two sets of loan processing costs.
	Locking in the interest rate for permanent financing in advance can require expensive hedges.
Escrow financing	Permanent financing is placed at the time of contract award, before construction.
	Proceeds of the loan are placed in an interest-bearing escrow account.
	ESCO draws down funds as needed, as in a short-term construction loan.
	Interest costs accrue on the permanent financing, but some of expense is recouped by interest earned on the escrow account.
	Often less expensive than two separate loans.

Source: U.S. Department of Energy 2005.

markets to ensure that the agency will receive the best possible overall value. The financing of an ESPC project is a contract between the ESCO and the financier, and it is the responsibility of the ESCO to obtain the competitive offers, evaluate them, and make a selection based on preestablished criteria for best value. The ESCO is required to document the selection of the financing offer in a certified Selection Memorandum that becomes part of the final proposal to the agency.

Over the past several years, efforts have been under way to reduce project financing costs. For example, the ESCO is now required to obtain at least three bids from potential project financiers. The total annual interest rate for the financing has two components:

- An index interest rate that depends mostly on the prevailing cost of money in the financial marketplace. Financiers of ESCO projects generally lock in the permanent financing the day before award of the delivery order. U.S. Treasury securities (5-, 7-, 10-, 20-, and 30-year) are commonly used as the index, as they have historically been stable indicators of the future cost of money.

- A premium fee to cover the lender's transaction costs, such as legal fees and administration. The premium on most super-ESCO loans has been in the range of 100 to 250 basis points (100 basis points equals 1 percent). The premium component may include hedging costs and any special project risks that the lender perceives, such as ESCO creditworthiness and track record, the technical complexity of the energy saving measures, the length of the contract, and so on.

Types of performance guarantees. At the heart of the ESPC is a guarantee of specified levels of cost savings and performance. The customer is not obligated to pay for an unmet guarantee and thus maintains a positive cash flow throughout the project period. The allocation of responsibilities between the agency and the ESCO defines the specifics of the guarantee—who does what and who pays for what during the term of the contract. The agency contracting officer must receive acceptable performance and payment bonds and any required insurance certificates before construction begins.

IDIQ contracts have a number of built-in protections for the agency, with the ESCO bearing more risk in a super-ESPC than in a conventional ESPC. Super-ESPC ESCOs have agreed to the terms and conditions of the IDIQ contracts, which specify that delivery orders issued against the super-ESPCs are not subject to protest procedures. Complaint resolution is handled by an ombudsman appointed by the head of the ordering agency and DOE.

The sharing of the savings between the agency and the ESCO is negotiated and defined in the contract, but the energy cost savings must be greater than the payments to the ESCO. The agency keeps any savings over the target. If the savings fall short of the target, then the ESCO's guarantee takes effect and covers the shortfall. The agency can also redirect performance payments to the financier should shortfalls occur. The contracts include requirements for the ESCO to repair or replace any malfunctioning or burned-out equipment during the contract period, to ensure that the savings continue to be achieved. At the end of the contract term, the title to the equipment is transferred to the agency, if it is not transferred at project acceptance, and the delivery order is terminated.

Measurement and Verification (M&V) procedures. To verify and document that the guaranteed savings are being delivered, the ESCO carries out an M&V plan that verifies the performance of the installed energy measures, quantifies the associated energy savings, and demonstrates proper maintenance. The M&V plan establishes the schedule for site inspections and specific measurements and monitoring, as well as the documentation that will be required for performance verification. All project-specific IDIQ task orders must have an M&V plan. The International Performance Measurement and Verification Protocol (IPMVP) process is used for documenting savings in ESPC projects.

If the actual annual savings, as determined by the M&V process, are less than the annual guaranteed savings amount, the ESCO must correct or resolve the situation or negotiate a change. The IDIQ contract specifies the guidance for reconciliation and the process for resolving disputes, which usually involve a third party. To date, however, DOE has participated in a very limited number of disputes.

During the M&V phase of the project, the number of measurements may decline over time as trends emerge that can reliably indicate future performance. The agency must approve any changes that the ESCO proposes to the M&V plan. Annual site inspections by the ESCO are mandated in the legislation authorizing ESPCs. The M&V costs for half of all the super-ESPC projects have been about 2.5 percent of first-year cost savings, which appears reasonable.

Policies to create incentives for public sector staff. An annual awards program has been established by FEMP that recognizes federal agencies for outstanding projects that contribute significantly to meeting federal energy and water saving goals. The annual recognition program encourages replication, innovation, and model projects, as well as providing outside validation of the projects to help agencies win support for increased project funding. The annual recognition program includes the category "Presidential Awards for Leadership in Federal Energy Management." Individual agencies also conduct their own employee award and recognition programs.

Several agencies include monetary awards that go to the agency department responsible for the energy savings, which may use the funds for agency approved, non-energy-related purposes. Some agencies, such as the Department of Agriculture, may also provide monetary awards directly to employees in the form of bonuses for measures or projects that result in significantly improved services, savings, and reduced energy consumption.

Project facilitators. To help agencies enter into ESPCs, DOE has established a FEMP ESPC team that can provide technical assistance and facilitate the process. These project facilitators guide the agencies throughout the ESPC process and consult with agency customers on contracting and financing issues, M&V, and technology and engineering issues. FEMP teams offer the assistance at no cost, starting with the kick-off meeting and extending through the review of an ESCO's initial proposal. Project facilitators are required for all super-ESPCs under the new, master IDIQ contract. Contractors retained by DOE provide this facilitation service, as DOE staff is limited. After the review of an ESCO's proposal, the support is available on a cost-reimbursable basis (usually $30,000 to $50,000) through system commissioning and the first year after project completion.

Acknowledgments

This case study was prepared by Brian Henderson. Information and materials provided by Schuyler "Skye" Schell, Brad Gustafson, David McAndrew, and William Raup of the Federal Energy Management Program, in the Office of Energy Efficiency and Renewable Energy at the U.S. DOE, were critical in its preparation.

Notes

1. UESCs may be growing in number in the future, as ESCOs team up with local utilities to implement energy efficiency and renewable energy projects in federal facilities. ESCOs are finding the sole-source UESCs a faster way to contract for projects without the new 2009 competitive requirements for ESPCs.

2. The National Energy Conservation Policy Act (NECPA) is the primary legislative authority directing federal agencies to improve energy management in their facilities and operations.

3. The Energy Policy Act of 1992 (EPACT), Public Law 102–486, amended NECPA to include additional energy management requirements and achieve target reductions.

4. On June 3, 1999, President Bill Clinton signed Executive Order 13123 promoting governmentwide energy efficiency and renewable energy.

5. This process may be affected in the near future by the passage of the National Defense Authorization Act, which is establishing more competitive procurement procedures for military and nonmilitary federal government contracts.

References

Efficiency Valuation Organization (EVO), "International Performance Measurement and Verification Protocol (IPMVP): Concepts and Options for Determining Energy and Water Savings Volume 1," EVO-10000-1.2007, http://www.evo-world.org.

Enabling documents can be found at http://www1.eere.energy.gov/femp/pdfs/uesc_enabling_docs_draft.pdf.

U.S. Department of Energy, Office of Energy Efficiency and Renewable Energy, *DOE Super ESPC Delivery Order Guidelines.* Version 3.06, April 2005.

—— Office of Energy Efficiency and Renewable Energy, *Federal Energy Management Program, Year in Review-2007,* http://www1.eere.energy.gov/femp/pdfs/annrep06.pdf.

—— Office of Energy Efficiency and Renewable Energy, *FEMP Newsletter,* Spring 2008, http://www1.eere.energy.gov/femp/pdfs/fempfocus_spring_2008.pdf.

—— Office of Energy Efficiency and Renewable Energy, *FEMP Practical Guide to Savings and Payments in Super ESPC Delivery Orders,* Revision 7, May 1, 2007.

—— Office of Energy Efficiency and Renewable Energy, *Generic IDIQ ESPC Contract,* http://www1.eere.energy.gov/femp/pdfs/generic_idiq_espc_contract.pdf.

—— Office of Energy Efficiency and Renewable Energy, *Introduction to Measurement and Verification for DOE Super ESPC Projects,* June 2007.

—— Office of Energy Efficiency and Renewable Energy, *Presidential Awards for Leadership in Federal Energy Management, 2009,* http://www1.eere.energy.gov/ femp/services/awards_presidential.html.

—— Office of Energy Efficiency and Renewable Energy, *Qualified List of Energy Service Companies: Evaluation Process,* http://www1.eere.energy.gov/femp/financing/espcs_qualifiedescos.html#ep.

—— Office of Energy Efficiency and Renewable Energy, *Super ESPC-Just the Facts,* FEMP fact sheet, 2008, http://www1.eere.energy.gov/femp/pdfs/espc_fact_sheet.pdf.

Canada—Federal Buildings Initiative Program

In 1991, the Canada Treasury Board authorized federal agencies to enter into multiyear energy performance contracts (EPCs) under a Federal Buildings Initiative (FBI), which was modified in 1995. In addition to helping the agencies to reduce energy use and costs, the FBI fits within the broader Canadian government framework of policies on environmental protection, fiscal responsibility, integrity of the contracting process, and international trade. The Treasury Board also authorized Natural Resources Canada (NRC) to develop documents and support services for the FBI to assist the agencies in pursuing EPCs. The FBI also provides other procurement and training services regarding energy efficient equipment for federal buildings. The Office of Energy Efficiency, which oversees the FBI within Natural Resources Canada, has a broad portfolio of energy efficiency programs, called ecoENERGY, for residential, commercial, and industrial customers.

The FBI is a voluntary initiative that offers the various departmental agencies a model framework for updating their facilities with energy-saving technologies and practices. The EPC provides a source of funds other than the agency's capital budget for carrying out energy efficiency projects. EPCs can also supply many other services, including energy analysis and audits, engineering and design, construction, commissioning, staff training, and maintaining and monitoring the performance of energy saving equipment after it has been installed. The FBI's energy management service contracts with prequalified, private sector ESCOs, also called "energy management firms" (EMF). Two types of EPCs participate in the program: first-out EPCs, in which the EMF retains 100 percent of the energy savings obtained either until the project is paid for or until the end of the contract, whichever occurs first; and shared savings EPCs, in which both the EMF and the agency receive an agreed-upon percentage or dollar amount of the savings over the life of the contract. Most government agency EPCs are of the first-out type.

To date, the FBI program includes over 85 retrofit projects, attracting Can$320 million in private sector financing, resulting in over $40 million in annual energy savings in more than 7,500 buildings across Canada. These FBI projects, representing 35 percent of agency building stock, have also helped to improve building comfort; promote a healthier, more productive workplace; and reduced the impact of operations on the environment, cutting greenhouse gas emissions by 285 kilotons. Energy intensity improvement has averaged more than 20 percent per project undertaken.

Performance of the energy audit. The FBI recognizes the preliminary energy audit or opportunity assessment as a critical first step in helping an agency locate the most appropriate actions to reduce energy use. It identifies the potential for energy efficiency improvement at the facility and the technical and financial aspects of a possible project, and can suggest ways to implement the measures identified. To assist the agencies, the FBI has provided guidelines for energy audits, including sample forms, worksheets, and graphs that can be used for data collection and as deliverables at specific audit levels. The preliminary energy audit is conducted before the energy management firm is selected.

The FBI has negotiated with local utilities to provide preliminary energy audits under the utility demand-side management or DSM programs. Some of the preliminary energy audits are free to the agencies, but cost penalties may apply if the agency does not implement the audit recommendations within a specific time period that the utilities establish. Also, an EMF may offer preliminary energy audits that are conditionally free. Here the firm offers to waive part or all of the charge for the service if the recommendations are implemented. However, the audit findings may indicate that moving forward with an EPC is not warranted, or that particular EMF may not be selected in the competitive selection process to implement the project. The agency must be prepared to cover the cost of the audit should this occur. The FBI and agencies have also used independent engineering consultants to perform the preliminary energy audit, as well as engineering staff from other agencies.

Energy audits may be provided at varying degrees of detail (or technical levels). The FBI has recognized four distinct types of energy audits—yardstick, screening, walk-through, and engineering. As noted in Table CS 2.1, they vary in scope, data required, complexity, deliverables, time to complete, and cost. The FBI requires two levels of energy audits: a preliminary screening audit and a detailed engineering audit.

Table CS 2.1   Summary of Audit Levels

Energy audit	Level	When to conduct	Days to complete	Approximate cost
Yardstick	Preliminary; minimum technical data and analysis, energy demand and use profiles, indication of potential	Basic data gathering to identify buildings that may have energy savings potential	1/2 to 1	$250 to $500
Screening	Preliminary; end use breakdowns, possible energy saving opportunities, preliminary savings estimates	Level of audit generally required to prepare an RFP for energy performance contracting (EPC)	1 to 3	$500 to $1,500
Walk-through	Preliminary; system type and equipment information, specific savings opportunities identified, preliminary costs and savings	Prior to bidding on RFP; conducted and paid for by prospective energy savings company (ESCO)	3 to 10	$1,500 to $5,000
Engineering	Detailed; extensive data gathering, modeling, simulation; leads to detailed implementation plan, drawings and specifications	ESCO that wins contract conducts detailed audit to develop implementation plan	10 to 50	$5,000 to $50,000

Source: Natural Resources Canada.

Types of energy saving projects. The FBI encourages a broad mix of energy efficiency measures. For example, the audits could cover the building as a whole or be very specific, covering certain systems or building operations. The FBI's Audit Standards Guidelines identified specific types of audits, such as lighting, electrical systems, mechanical systems, operations, and maintenance, while also addressing air quality, waste management, and environmental issues. The EMF and the agency also might agree to implement a project in a variety of ways. In some EMF-agency arrangements, the approach is to conduct major studies encompassing all possible improvements before proceeding with any project—a single path approach; others may study and present measures individually, implementing them as they are approved—a multiple path approach. Discussions with FBI staff have indicated an interest in establishing a regional ESCO approach, similar to the U.S. Federal Energy Management Program's super-ESPC contracts, such that one competitively selected ESCO might service an entire region of the country to expedite the ESCO selection and contracting process.

Evaluation criteria used to select the ESCO. To choose an ESCO, an agency must use a tender bidding process. To make the process simpler for the agency, the FBI has created a qualified bidders list of prescreened EMFs capable of carrying out energy performance contracts. Currently, there are 10 EMFs on the qualified bidders list, and they remain on the list for two years before being required to reapply. The FBI publication "Qualification Requirements" lists 11 criteria for selecting an EMF (see Table CS 2.2).

Public agencies are generally required to issue the model RFP to the list of qualified bidders to select an EMF for a specific project. The model documents of the FBI are often customized to suit the needs of a particular project and the specific agency. The proposed contract is included with the RFP, so that all the terms and conditions of the contract are seen before tenders or proposals are submitted. The RFP includes a copy of the preliminary energy audit, along with recent studies, any specific retrofits to be included or excluded, special operating requirements, the target payback period, and any expectations regarding sharing of risk.

The FBI model RFP includes a suggested scoring method for evaluating information supplied in proposals. An important consideration is the amount of emphasis to place on the details supplied by bidders on the retrofits that they propose. The evaluation process anticipates that the proposals from different EMFs will be quite different, both in the savings and cost they propose and in the type of improvements proposed, recognizing

Table CS 2.2   Selection Criteria for the Qualified Bidders List

Criteria	Minimum requirements
Project experience	Has completed at least three recent energy improvement projects
Contract capability	Demonstrates recent financial performance and access to multiyear financing resources
Engineering design experience	Demonstrates capability of designing and installing energy efficiency measures and other required services
Human resources	Has personnel with at least two years' experience in design and implementation of energy improvements
Project management approach	Indicates how project is to be managed
Workers' compensation certification	Demonstrates compliance with workers' compensation claims
Financial capability	Provides financial statements for most recent three years
Contract duration	Has acceptable contract duration
Risk management	Has evidence of adequate insurance coverage and can comply with bonding requirements, if required
Partnership and joint ventures	Can identify each affiliated, parent, or subsidiary company
Dispute settlement approach	Uses resolution method for all engineering, subcontractor, and supplier disputes

Source: Natural Resources Canada 2008.

that no textbook method exists for determining appropriate measures or for estimating savings.

For large facilities with multiple buildings, bidders may be asked to prepare proposals on a portion of the project, but still giving the agency enough insight into the type of improvements a bidder will implement throughout, while reducing the bidder's proposal preparation time and costs. In this case, the EMFs will be invited to bid on a representative building or group of buildings, which have good replicability across the entire facility.

A committee of agency representatives and possibly other agency designees (Office of Energy Efficiency, Public Works and Government Services, etc.) reviews and ranks each of the proposals. The model RFP includes sample criteria and a scoring method, with the relative weighting of the criteria to be customized by the agencies for their respective projects. A sample scoring sheet is presented in Table CS 2.3.

The top-ranked EMFs (based on the highest-scoring two or three proposals) are invited to make an oral presentation to the agency's selection committee on their respective approaches to the project. A combination of the proposal score and the oral interview determines the firm selected for the project.

Table CS 2.3   Example of Weighting the Criteria in Proposal Evaluation

Category	Maximum point value	Weighting factor
Financial: payback period; cost breakdown	100 points	25% (0.25)
Technical: completeness of energy savings estimate; engineering approach	100 points	25% (0.25)
Implementation: plan for making improvements; monitoring savings	100 points	20% (0.20)
Operation and Maintenance: preventive maintenance approach	100 points	10% (0.10)
Project Management: qualifications of personnel	100 points	10% (0.10)
Training: Approach for delivering training	100 points	10% (0.10)
TOTAL SCORE	–	–

Source: Natural Resources Canada 1995.

Multiyear energy performance contracts. The RFP requires the EMFs to propose multiyear financing projects. Initially, the program had an eight-year cap on the length of an EPC. However, the duration can now be based on a cost-benefit analysis, with terms up to 15 years. A first-out EPC is used, under which the EMF retains 100 percent of the energy savings until it recovers all of its costs, including profits, or until the contract term is over, whichever occurs first. The contracts also require the submission of a performance bond by the EMF when the energy efficiency improvements to be implemented are agreed on. This performance bond remains in effect and is subject to forfeiture for nonperformance until the equipment and systems are accepted by the agency.

Project financing. The FBI program is designed for the EMFs to provide 100 percent of the necessary project financing. The RFP requires the EMF to clearly define the source of the financing, interest rates, term, payment flexibility, and any other financing-related obligations. Private sector project financing is widely available to Canadian EMFs at competitive rates, from local banks and specialized divisions of international financing companies. Neither the EMF nor any financial institution may retain title to any equipment purchased or installed as part of the improvements as security for the financing. To reduce the amount of project financing, the EMFs have also secured ratepayer-funded energy efficiency monies, where available, from local utilities under their DSM programs.

Types of performance guarantees. Payments to the EMF are based solely on the energy (and sometimes water) savings realized from the energy efficiency project. The EPC contract includes a savings or performance guarantee over the life of the agreement. If the full efficiency improvement does not occur, then the EMF must cover any shortfall in savings. This contract also assures the agency that the EMF will have no claim beyond the contracted savings. The savings guarantee must be sufficient to repay the EMF's investment in the project over the contract term. The guarantee also provides some further level of assurance that the energy savings should continue to accrue after the EPC ends, through the useful life of the equipment or other measure.

M&V procedures. The frequency of monitoring reports from the EMF depends on the stage of the project. During the initial design and construction phase, monthly reports are usually required. Generally during the first two years of postconstruction monitoring, quarterly reports are needed, and after that, semiannual reports are provided in most cases. Some agencies with more complex projects require monthly reports during the M&V phase on the performance of the energy efficiency improvements, energy and operating cost savings, variances (weather, change in facility operations, etc.), and applicable corrective actions.

The energy performance contracts contain sections on arbitration procedures. The EMF and the agency are expected to make all reasonable efforts to resolve any dispute, controversy or claim arising out of the project through discussions and good faith negotiations. In event of an unresolved disagreement, the matter goes to arbitration before an independent panel of three arbitrators, whose determination is final and binding on both parties.

Providing incentives to public sector staff. The agencies have considerable freedom to adapt implementation and contractual documents to their particular requirements, as long as the result produces energy savings. The contracts also encourage further savings, as any increase in savings will retire the EMF's investment more quickly, and the agency will be reaping 100 percent of the savings sooner (cost savings flow back to the agency's operating budget). No direct financial incentives or commissions are provided to public staff associated with their work on EPC projects. However, some agencies do acknowledge exceptional efforts in championing energy efficiency, and individuals may receive annual performance or merit awards of several hundred dollars, in addition to the traditional employee recognition awards.

Project facilitators. The Office of Energy Efficiency manages the FBI program. The FBI offers a number of services and publications to help the agencies. In addition to prequalifying energy service companies, the FBI produces sample tender documents, as well as the *FBI Update Newsletter,* which highlights success stories and case studies from the federal agencies and best practices related to energy management. The agency Public Works and Government Services Canada also offers services to assist the federal government departments in managing their EPC projects. Private consultants are also available to guide agencies' procurement and management activity.

Acknowledgments

This case study was prepared by Brian Henderson. The information and materials provided by Phil Jago and George Izsak, of the Office of Energy Efficiency, Natural Resources Canada, and their consultant Jim Gilroy, of Goss Gilroy, Inc., helped in its preparation.

References

FBI contacting arrangements at http://www.oee.nrcan.gc.ca/publications/fbi/m92-201-2000e-3.cfm?attr=28.

Natural Resources Canada; model RFP documents are available from the Federal Buildings Initiative, *FBI Model Documents*, Ottawa, http://oee.nrcan.gc.ca/ publications/fbi/pdf/2701473E.pdf.

——— Office of Energy Efficiency, *FBI Implementation Documents*, Report M27-01-473E, Ottawa, July 1995, http://oee.nrcan.gc.ca/publications/fbi/pdf/2701473E.pdf.

——— Office of Energy Efficiency, *Federal Buildings Initiative: FBI Implementation Documents*, Ottawa, 1995.

——— Office of Energy Efficiency, *Federal Buildings Initiative—Qualification Requirements*, Ottawa, 2008.

——— Office of Energy Efficiency, *Federal Buildings Initiative: The Policy Context*, Ottawa, 1995.

———— Office of Energy Efficiency, *Federal Buildings Initiative*, Ottawa, 2008, http://www.oee.nrcan.gc.ca/communities-government/buildings/federal/federal-buildings-initiative.cfm.

———— Office of Energy Efficiency, *The Federal Buildings Initiative: Audit Standards Guidelines*, Ottawa, 2002.

———— Office of Energy Efficiency, *The Federal Buildings Initiative: Audit Standards Guidelines*, Ottawa, http://oee.nrcan-rncan.gc.ca/publications/fbi/m92-246-2002e-4 .cfm?attr=28.

———— Office of Energy Efficiency, *The Federal Buildings Initiative: Managing Energy Performance Contracts in Federal Buildings*, Ottawa, 1994.

———— Office of Energy Efficiency, *The State of Energy Efficiency in Canada, Report 2006*, Ottawa.

Energy Performance Contracting in France

The concepts of energy services and performance contracting evolved in France many decades ago. Since the 1940s, a number of private companies have offered services under the name *exploitation de chauffage*, now referred to as "energy performance contracting" (EPC). These companies were established to operate the heating, ventilating, and air conditioning (HVAC) equipment in buildings through operational contracts with the building owners; these were multiple-year contracts and provided some guarantee of results. The contracts included the aggregation of financing, operation, and guaranteed savings, or ownership and operation of the energy equipment, with a fixed price for the energy services delivered to the customer.

Operational contracts for public and semipublic buildings within public purchase rules led to a coding of these contracts to ensure the indexing of their prices, to apply differentiated rates of the value added tax (VAT), to distribute the elements of the invoice in accordance with the law between owner and tenants or occupants, and to enter them in the public accounts. This coding and the requirement for public accounting for fixed results for a fixed price largely determined the features of traditional energy performance contracts in France.

EPC models. There are various energy services models in France but they derive largely from a single template. The dominant model is service based, wherein the various components are broken out (can be taken or not by the user) and the energy service provider (ESP) is judged (and paid) on each of the separate components, for example, economic efficiency in purchasing energy, technical efficiency in continuous audit and maintenance, financial efficiency in planning works on time, technical and economic efficiency in proposing energy saving measures that can be repaid from benefits, and so on. A lengthy operations and maintenance (O&M) contract is part of the deal and provides a real incentive for long-term performance.

The traditional French operating contracts involve heating and air conditioning installations for which the service provider has a firm commitment, for example, undertaking to guarantee a temperature level for the heating of premises. The ESP is responsible for supplying the resources necessary to achieve a specific result. Some key characteristics of these contracts are the following:

- The contracts define an obligation for results; usually for reliability and lower energy consumption.
- The contracting party provides diagnoses and carries out the improvements to ensure the economic benefits.
- Contract duration typically ranges from 6 to 12 years.

Types of services. The operating companies offer a wide range of energy services, including the management and reporting of energy use; management; O&M of energy equipment; replacement of equipment, including installation and financing, and so on. These services can be broadly categorized as shown in Table CS 3.1.

Table CS 3.1   Types of Services Offered by Operating Companies

Energy supply	Decision making	Equipment supply	Installation	Operations	Funding
Network access Consumption monitoring and billing Quality and safety	Advice Energy audits Feasibility studies Project management	After-sales service Performance guarantee	Design Equipment selection and installation	Maintenance Control of installation Equipment performance guarantees	Loans Leasing Third-party Financing

Source: Agency for the Environment and Energy Management (ADEME) 2006.

French EPCs have been classified into four types, as shown in Figure CS 3.1.[1]

Within the four main contract types are many variations having to do with fuel prices, escalation clauses, climate adjustment, and so on.

Barriers to EPCs. The barriers to the implementation of energy services and performance contracting in the public sector include the following:

- *Political*
  Energy is a small part of the public authority budget and does not get the attention of decision makers.
  The short political horizons of decision makers preclude the approval of energy efficiency measures with paybacks longer than a few years. Energy savings measures' lack of visibility makes them politically unattractive.
- *Organizational and Institutional*
  Lack of technical capacity
  Limited resources
  Separation of capital and operating budgets
  Annual budget cycles
- *Financial*
  Limited internal funds availability
  Lack of expertise concerning financing options

Figure CS 3.1   The Four Contract Types

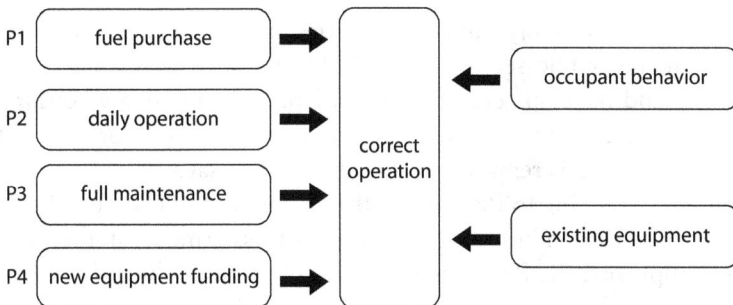

Source: Adnot et al., 2002.

- *Legal*
  Difficulties in combining goods and services in a single contract
  Complexity of EPC contracting

As a result of these barriers, and the fact that French laws have precluded investment by the private sector in public facilities, the *chauffage* model has not been applicable in public energy efficiency projects. But a new law on public-private partnerships enacted in 2004 greatly facilitates the use of ESPCs in the public sector.

Introduction of public-private partnerships. The French government recently introduced the concept of public-private partnerships (PPP), which have changed the investment funding context for the public sector.[2] These special contractual arrangements in effect allow performance targets to be introduced into invitations to tender, particularly with regard to the level of energy consumption to be attained. The draft law on the implementation of the Grenelle Summit on the Environment provides in Article 5 that "the thermal renovation work on the 50 million square meters of state buildings and 70 million square meters of public establishments will be made using as a priority contracts for PPPs, including energy performance contracting." (ADEME, 2008, p. 5).

PPP contracts are global administrative contracts by which a local authority or a public establishment can contract with a private organization for the funding, design, production, conversion, operation, or maintenance of public equipment or for the funding and management of services. This type of contract is not a public contract or the delegation of a public service, but represents a new category of contract that has a specific award procedure, even though there are numerous similarities to the delegation of a public service.

The PPP is for many public authorities an ideal way to contract for savings in the public sector, since in EPC the energy service provider identifies and develops energy efficiency projects in energy-consuming sites; finances them; most often operates or maintains all or part of the new facilities; and is remunerated on the basis of savings.

The order relating to PPP contracts emphasizes the need for the public authority to conduct and document an assessment of the project before employing the PPP procedures. In this process, the public authority has to justify either the complexity of a project or its urgent nature, documenting the economic, financial, legal, and administrative reasons that led to the selection of the PPP contracting process. As a part of this process, a comparative analysis of the overall cost, performance, and risk sharing of the various options must be conducted.

Baseline audit. Prior to the initiation of the contracting process, the public authority is likely to collect baseline information, identify the potential energy savings measures and options, and define the needs and expectations of the project. Depending on the technical expertise of in-house staff, the agency may engage consultants in this effort. Although a full investment grade audit is not performed at this stage, preliminary planning and feasibility assessments may be included to develop the project scope and to support preparation of a tender notice (*avis d'appel public à la concurrence*, or "AAPC"). The agency may also define the nature of the "competitive dialogue" to be conducted, identify the candidates to be invited for the competitive dialogue, and specify the criteria for evaluation and selection of the winning contractor. The public authority may also establish the organizational structure and procedures for the competitive dialogue.

The competitive dialogue. The procedure for placing contracts (except in the case of an emergency) is the use of a procedure called the "competitive dialogue," in which the public authority engages in an informal process of soliciting and discussing proposals from various service providers. The competitive dialogue process is specified in the PPP procedures for awarding public works contracts, supplies, and services. The dialogue procedure is used whenever the public authority wants to attract private initiatives in technical, legal, or financial areas. The authority identifies candidates that may be able to provide the needed services and will then undertake discussions with operator candidates on the key elements of the project and the economic model. The order also specifies mandatory award criteria and provides a non-exhaustive list of them, such as performance targets, overall cost of the proposal, its innovations, and so on. The competitive dialogue is more suitable for complex projects involving multiple energy savings measures than for simple projects that can be easily specified.

The competitive dialogue process is generally initiated with a notice of public competition (the AAPC), which provides a description of the nature of the project, its objectives, and expected performance. Subject to these specifications, the dialogue can cover various options related to the technical, legal, and financial aspects of the project. This process allows the ESPs to develop innovative approaches and provides the public authority a range of options.

After a review of the discussions during the dialogue, the public authority identifies the options and solutions that meet its needs, and then it invites the candidates still in contention to submit their final offers on the basis of the solution or solutions they presented and discussed during the dialogue (the solutions of different candidates may not be the same). These final offers are typically made in a period of about one month and include a detailed presentation of the technical offer, services to be provided, and the financial bid.

This phase of the proceedings is certainly the most difficult. Indeed, the process stipulates that "the number of candidates should be sufficient to ensure genuine competition."[3] The public authority therefore needs to maintain genuine competition up to this stage, while using elements from the dialogue, without violating the protection of business or preferring a particular candidate in the race. Each candidate still in contention takes into account any information provided by the public agency and makes an offer based on its own solution, consistent with its expertise. A key challenge in assessment is that the competing offers may be quite different from each other and may include some unique and possibly proprietary features.

Contract award. The contract is awarded to the candidate who presented "the most economically advantageous" proposal, on the basis of, "for example, quality, price, technical merit, aesthetic and functionality, environmental characteristics, cost effectiveness, after-sales services and technical assistance, [and] the delivery date."[4] The ordinance provides for a weighting of the criteria, or if that is "objectively impossible," a hierarchy of the criteria.

The contract duration is based on the following considerations:

- The period of repayment of the investment and maintenance cycles. (The duration of the contract should be as close as possible to the life of the asset and its components, to optimize synergies between construction and maintenance.)
- The time required to absorb the funding of the project and to get a fair return on equity invested, calculated according to a rate of return determined at the beginning.
- The duration of optimal sharing of benefits between the agency and the contractor.

Performance guarantees. Two types of performance guarantees may be addressed concurrently or substituted, depending on the nature of the project:

- Technical performance, measured in terms of quality of service to the public (including equipment availability).
- Performance, in terms of the frequency of works and related services under the principle of shared risk.

Compensation for unsuccessful candidates. A unique feature of the French PPP process is the compensation of the unsuccessful participants in the competitive dialogue. While the competitive dialogue procedure does not require compensation of the unsuccessful candidates, the public authority may elect, depending on the level of services requested and the nature and intensity of the dialogue, to provide compensation for a portion of the costs incurred by the unsuccessful candidates. In particular, compensation is provided (a) when it allows more candidates to participate, thereby increasing competition; (b) when it improves the chances of obtaining better-quality offers; or (c) when it reduces the risk of litigation regarding the contract award.

The compensation rules and procedures will be clearly specified in advance so that the participants in the competitive dialogue process fully understand what costs are likely to be reimbursed.

PPP timetable. The time required to complete the PPP process depends on the size, type and complexity of the project. A typical timetable is as follows:

Preparatory stage
- From AAPC to prequalification submissions—1.5 months
- Confirmation of prequalification consortia—1 to 2 months

Competitive dialogue
- From confirmation of prequalification to initiation of dialogues—1 to 3 months
- Completion of competitive dialogues—4 to 12 months

Final stage
- From completion of competitive dialogues to submission of final offers—2 to 4 months
- Clarification/negotiation and contract execution—2 to 6 months

Acknowledgments

This case study was prepared by Dilip Limaye. Valuable information and clarifications for the development of the study were provided by Dr. Frederic Rosenstein, of the French Agency for the Environment and Energy Management (ADEME), and research support by Lucas Bossard.

Notes

1. For more information on French ESPs, see Adnot, Guerre, and Jamet, *A Short History of Energy Services in France*.
2. Article 11 of Order no. 2004-559, June 17, 2004.
3. Directive 2004/18, March 31, 2004.
4. Article 29 of Directive No. 2004/18/EC, March 31, 2004.

References

ADEME, 2008. "Services énergétiques et Contrats de performance énergétique: des outils pour la mise en œuvre de Grenelle." ADEME & Vous: Stratégie & études, No. 14, September 2008. [The full article can be found at http://www2.ademe.fr/servlet/getBin?name=9F6E58FDDA7A0BCF4B4BC D4780E983F31221117212987.pdf].

Jerome Adnot, Florian Guerre, and Bernard Jamet, *A Short History of Energy Services in France*, 2002, http://www-cep.ensmp.fr/english/themes/mde/ pdf%20J%20Adnot/pef9.pdf.

M. Dupont and J. Adnot, 2004. "Investigation of Actual Energy Efficiency Content of 'Energy Services' in France," in *Proceedings of the International Conference on Improving Electricity Efficiency in Commercial Buildings*, Frankfurt, April 2004.

Paolo Bertoldi and Silvia Rezessy, *Energy Services Companies in Europe*, European Commission, Joint Research Centre, 2005.

Simon Ratledge and Paul Lignieres, Linklaters 2006; *PPP in France* 2006.

Energy Performance Contracting in Germany

Germany has a relatively long history of energy performance contracting (EPC) in the public sector, dating back to the early 1990s, and thus is considered by many the European country with the most mature energy service company (ESCO) industry (Seefeldt 2003; Brand and Geissler 2003). By 2005, the contracting market in Germany had a total investment volume of about €5 billion. The vast majority, 85 percent, is for energy supply contracting (ESC), mostly for heat; only about 10 percent for EPC; and the remainder for other forms such as energy operations contracting (EOC).[1] The different types of contracting are outlined in Figure CS 4.1.

The supply side of the market is characterized by a large number of ESCOs, about 500, most of which are providing ESC services, mostly for heating, and energy operation contracting. Although about 50 ESCOs have some experience with energy performance contracts, only about 20 have more than one reference project.[2] The background of ESCOs is fairly diverse, ranging from subsidiaries of large utilities and former municipal utilities to equipment suppliers, construction companies, and engineering and consulting firms. ESCOs are active in both the public and private sectors. Two ESCO associations exist, with membership overlapping significantly.[3]

Figure CS 4.1   Types of Energy Contracting

	Energy Supply Contracting (ESC)	Energy Performance Contracting (EPC)	Energy Operation Contracting (EOC)
client	energy consumer	user of existing units/equipment	user and owner of existing units/equipment
target	energy supply	realization of energy saving potentials	economically optimized operation
services by contractor	planning, construction, operation, maintenance, financing	partly renewal (financing incl.), operation & maintenance	operation
refinancing	energy sales	energy savings	operation fee
risks borne by contractor	risks of construction, operation, maintenance, finance, and purchase	risks of actual energy savings, of O & M	risks of maintenance and replacement investment
economic advantages for client	avoided investment, purchase/ bulk buying advantages, reallocation of risks	energy savings guaranteed by contract	technical optimization and professional experience of contractor

Source: Kuhn 2006.

Energy efficiency in the public sector. In this case study, the definition of the public sector was narrow, comprising only facilities that are owned and managed by federal, state, or local authorities or municipal governments. It excluded the large number of facilities owned by government-owned corporations and similar entities (Prognos 2006a). One reason for using the narrow definition is that only in these facilities can the public sector actually influence energy consumption. According to a recent study of the energy efficiency improvement potential in Germany, the energy consumption of the public sector is about 236 Petajoules, accounting for only 2.5 percent of total energy consumption in 2002 (Prognos 2006a). Under the narrow definition, public sector buildings include technical equipment (mostly HVAC and lighting) that accounts for about half of consumption. The other half is attributable to office equipment (such as computers) and public lighting. The technical and economic energy efficiency improvement potential in the public sector is about 20 percent of 2002 energy consumption.

For many reasons, however, the German public sector has been reluctant to engage in energy saving behavior and investment, and the potential remains unrealized. The main barriers include the following:

- Tight public budgets at all levels of government, preventing investments to replace and modernize equipment

- Separation of budgets for capital investment and operations
- Lack of incentives to save energy
- Lack of technical knowledge to optimize operations
- Inadequate or nonexistent energy management functions or capabilities
- Organizational complexity, with different organizations and agencies responsible for different aspects of energy use, such as users, technical supervision, investment, operations, and so on

Since 2000, climate change has become an important driver of energy efficiency policies and programs in Germany. As in most European Union (EU) member countries, national and subnational policies regarding energy efficiency are largely driven by EU energy and climate change policies, such as the European Climate Change Programs of 2000 and 2005. In early 2008, the EU committed to a 20 percent reduction in greenhouse gas emissions, to a 20 percent share of renewable energy in final energy consumption, and to a 20 percent reduction in energy demand by 2020, compared to 1990 levels. The 2006 EU end use efficiency and energy services directive specifies the reduction of final energy use by 9 percent by 2016. Under the national climate protection program of 2000, the German government has made a commitment to reduce the CO_2 emissions of its own properties by 30 percent (recently increased to 40 percent) over the period 2008 to 2012, compared to the reference year of 1990. This is to be achieved by its own investments, for which an annual budget of €120 million for the period 2006–2011 has been provided.[4] The other instrument to be used is energy contracting, based on earlier successful local and state experiences.

EPCs in the public sector. An estimate was recently made of the potential for energy contracting in the public sector, based on the properties owned by federal, state, and local authorities. The total number of properties is more than 185,000, most of them owned by local governments. More than 80 percent of energy consumption and about two-thirds of total energy cost in public sector buildings is for heat, with the remainder for electricity (Prognos 2006b). The estimate of total potential for energy contracting in the public sector takes into account that contracting can only be done in buildings with their own heating systems (not in buildings whose heat is supplied by the district) and buildings large enough to minimize the transaction costs of contracting. Given, furthermore, that some barriers will not be completely removed in the midterm, the potential for contracting that could be realized by 2016 includes only about 20,000 properties.

But their annual energy costs of €1,070 million account for about 30 percent of overall energy costs. Energy costs savings of 20 to 30 percent could be realized with contracting, resulting in annual energy cost savings of €200–300 million for the public sector. Of this overall potential, contracting has already taken place since 1990 in about 2,000 properties, with annual energy costs of €150 million (15 percent), more or less equally divided between supply contracting and performance contracting. Thus the remaining potential over 10 years (2007–2016) is about 85 percent of the total potential, consisting of about 18,000 properties with annual energy costs of about €900 million, of which up to €300 million could be saved by contracting (Prognos 2006b).

There are many reasons that EPC was slow in garnering market share and why it still is not being done to the extent possible. Table CS 4.1 lists the most important barriers, solutions, and their status as regards EPC in Germany.

Table CS 4.1   Barriers to Energy Performance Contracting in Germany

Barriers to EPC	Solutions	Status
Uncertainties regarding procurement and budgetary requirements of EPC	Develops guidelines for procurement and budgetary requirements; follows the essential requirements of competition and comparison with an internal solution	EPC that is allowed under budgetary laws and under German/EU procurement law; remaining uncertainties that are due to differing requirements in different German states
Lack of information, awareness, and knowledge about EPC	Has more and better energy management and information/consulting; support through an energy agency	Ongoing effort
Lack of knowledge/ competence to deal with procurement complexity, e.g., economic comparison of different offers	Has support from external experts; development of guidelines and model contracts	Ongoing effort
Lack of incentives for staff	Creates incentives, such as recognition, prizes, etc	Little change; still a barrier
Technical staff fear of losing control and/or jobs	Uses different job description regarding supervision of the contractor and project control	Still relevant barrier

(continued)

Table CS 4.1   Barriers to Energy Performance Contracting in Germany *(continued)*

Barriers to EPC	Solutions	Status
Laying off public employees no longer required under EPC can be difficult; requirements that contractors use such staff create disincentives for EPCs	Has no easy solution; requires changes in public sector employment regulations	Still relevant
Transaction costs of EPCs	Uses standard documents (procurement, contracts) and guidelines; bundling of EPC projects across several smaller municipalities	Ongoing effort

Source: Author.

Without policy support at the national, state, and local levels to eliminate the barriers, energy performance contracting would not have been possible in public properties. Over the past 10 years or so, environmental and climate change concerns have led many governments at all levels to adopt sustainability strategies and programs, including promotion of renewable energy and energy saving measures, particularly in their own facilities. EPC is considered an essential instrument to fulfill the energy efficiency commitment and is part of the climate strategies of the national and state governments and many municipalities. Major milestones of policies influencing the acceptance of EPC are listed in Table CS 4.2.

In Germany, first experiences with EPC in the public sector began in the early 1990s,[5] with somewhat broader efforts taking off in the late 1990s, when EU energy markets were liberalized and the Maastricht treaty went into effect. The treaty requires the convergence of member states' economic policies and strict fiscal discipline at all levels of government. Energy agencies were set up in some states and worked with state and local governments to ensure EPC compliance with procurement and budgetary rules; public entities were allowed to enter into performance contracts with ESCOs and to develop standard documents to enhance the transparency of procedures. Many cities, required to reduce their operational costs but unable to fund the rehabilitation of their facilities from their own budgets, began entering into contracts with ESCOs. Independent experts often managed project development and tender procedures.

Table CS 4.2   Milestones of Measures to Support Energy Performance Contracting (EPC) in the Public Sector

Year	Milestone
1991	Maastricht Treaty created the European Union (EU) and set in train the process of economic and monetary union.
1992	Berlin Energy Agency (BEA) established by the State of Berlin in public-private partnership with utilities and banks.
1994	Berlin Government Energy Concept to reduce CO_2 emissions by 25 percent in 2010, compared to 1990, with special emphasis on energy savings in public sector facilities.
1996	First two model EPC pools in Berlin developed by BEA.
1998	Hessen EPC guidelines. UBA EPC guidelines.
1999	Federal EPC guidelines.
2000, 2005	European Climate Change Program. German National Climate Protection Program: commitment to reduce greenhouse gas emissions in federal government–owned facilities with innovative measures, including contracting.
2001	DENA, the German Energy Agency, starts operations.
2002	Pilot project, "Energy Contracting in Federal Properties" (with DENA).
2003	DIN Norm Contracting 8930 Part 5 is published, defining contracting and its various types, forms of remuneration, conditions of use, legal foundations, etc.
2003/2008	Federal EPC Guideline (DENA).
2006	EU end use efficiency and energy services directive: public sector to take a leading role in energy efficiency and support EPC.
2008	EU 20-20-20 commitment.

Source: Author, based on information cited elsewhere in this case study.

As government commitment grew, the project Energy Efficiency Contracting in Federal Properties was started in 2002 to further government energy goals using innovative solutions that do not require public financing.[6] The basic procurement and public budgeting barriers were dealt with in Germany fairly early on, state by state and in the federal government. It is now generally acknowledged that EPCs are allowed, provided that procurement is competitive and that the comparison of offers includes self-implementation of projects (i.e., without ESCO involvement). Another main factor has been the support that public clients receive from energy agencies. This takes the form of guidelines for gathering baseline information, model procurement contracts and guidelines, and support during the procurement process and in the commissioning and postimplementation phases.

Dealing with budgeting and EPC procurement issues. The major questions that initially prevented EPC from gaining acceptance among public sector clients were budget and procurement related:

- Is it permissible to finance public investments through energy cost savings (i.e., EPCs), in effect paying for them from the budget for current expenditures?
- If investments are financed by a third party under EPC, will this count against the debt limit?
- Under what conditions will EPC in locally owned facilities be allowed by supervising authorities at higher levels?
- How are payments to contractors to be budgeted?
- How will payments that occur in the future be handled?
- Is EPC compatible with the German procurement law (and EU regulations), that is, can existing procurement regulations be used? Specifically, can goods, works, and services be procured together in one contract?

The energy agencies in several German states, notably Berlin and Hessen, supported the development of demonstration projects starting in the mid-1990s. Clarifications about the legality of EPC in the public sector, with respect to budgetary and procurement laws, were sought and obtained. They were incorporated into guidelines and model contracts made available by several state and federal institutions, for example, the state of Hessen, the Federal Ministry of Transport and Construction (BMVBW), and the Federal Environment Agency (Umweltbundesamt - UBA).[7] Written opinions by relevant ministries, for example, in the State of North Rhine–Westphalia (NRW), confirmed that contracting conforms to procurement and budgetary laws. The latest guideline applying to contracting in federal properties is the "Leitfaden Energiespar–Contracting" of 2008 (DENA 2008).

Clarification of budgetary issues. All public properties in Germany, whether federal, state, or municipal, can in principle use EPC as an alternative to credit financing. As long as measures carried out under energy performance contracts are profitable—cost savings are greater than payments for contractor services—they are not taken into account for debt limits in most German states. Municipalities must have all credit transactions, and transactions similar to credits, authorized by the supervising authorities ("Kommunalaufsicht") except those for operations and maintenance

(O&M). Energy performance contracts emphasize a service of the contractor, rather than the investment, and can be considered to cover operating expenses if their term does not exceed 10 years. If the contract includes a guarantee of the profitability of the measures to be taken, and if it has been awarded competitively, no preauthorization is required. Payments to the contractor must be budgeted as building maintenance expenditures. In the case of the monthly advance payments to the ESCO, only about 80 percent of the ESCO share in the guaranteed savings is paid, in case actual savings are less, inasmuch as budgeting reimbursements would be difficult.[8]

Some uncertainty remains because supervisory authorities in different German states may hold different views on how to handle EPC, resulting in diverging practices in approving contracts. The most liberal state regarding EPC is North Rhine–Westphalia, where local authorities do not need a separate authorization for each EPC. In six other states each EPC has to be authorized separately, and its value is usually counted against the debt limit of the municipality. The remaining states fall somewhere between those extremes. In the city-states Berlin, Bremen, and Hamburg, there is no separate supervising authority.[9]

For federal properties entering into long-term energy performance contracts with total annual payments to the contractor (out of energy savings) above €300,000 (or if the investment aspect is more relevant than the service aspect), it is necessary to include a commitment appropriation ("Verpflichtungsermächtigung"), that is, an authorization for financial commitments beyond the current fiscal year, into the federal budget for the future payments to the contractor, generally before the tendering process begins. Furthermore, it is strongly recommended—and required for federal properties—that they compare the costs and benefits of investments under an EPC with those of such a project implemented on their own ("Eigenregie").

While the legality of EPC has been clarified, some claim that public sector energy efficiency initiatives may be hampered by unfavorable budgeting principles, namely, the separation of investment and operation budgets, which does not permit applying energy cost savings to new energy efficiency investments (e.g., WEC 2008). EPC that includes financing should be a way around this problem, but in any case, it is expected that it will disappear after completion of the ongoing reform of municipal budgetary law in Germany.[10]

Some segments in the public sector have funding arrangements that may be better suited to carrying out energy efficiency investments through EPC, and in fact provide incentives to do so. The public hospital sector seems to be a case in point. Under its dual financing system, invest-ment is financed by the state, whereas operating costs are covered by patients and their health care insurance companies in the form of flat rates per patient. Reduction of energy costs through investment thus ben-efits the hospital directly and frees up funds for other purposes.[11]

Clarification of EPC public procurement issues. German public procure-ment law allows EPC, but its implementation is complex and time-consuming. The German law against limitation of competition requires public agencies to award contracts in a competitive process; to secure equal treatment of all bidders; and to facilitate participation by small and medium-sized enterprises through division into lots. Contracts must be awarded to competent bidders, excluding nonrelevant considerations, and the award must be based on the most economical offer.

The law also specifies that small and medium-sized bidders should be enabled to bid for public contracts by separating bids for goods, works, and services, unless there are special technical and economic reasons to com-bine them. For EPCs such combination is indeed essential, and therefore the different components can be procured together according to German procurement regulations (see, for example, *Hessen Leitfaden* 1998/2003).

The German public procurement regulation provides two basic kinds of contracts and procurement processes that are applicable to EPCs. Depending on the focus of the services to be carried out, either a construction-type (VOB) contract or a service contract (VOL) is sought. Larger procurement has to be tendered within the EU, but the thresholds requiring EU-wide procurement are different for the two types of contracts: above €0.211 million for VOL and above €5.278 million for VOB. Procurement is typically EU-wide, even if below the thresholds. For EPC projects for federal properties, VOB is usually applied. In other juris-dictions, VOL is justified with the argument that requirements can be described only functionally, that is, that the objective of the contract is the percentage reduction of energy costs. The basic procurement process and rules are essentially the same for the two types of contracts. Which one to choose is often a political decision, since different agencies may be respon-sible for supervising the process.

German law and EU directives specify several types of procurement, ranging from open, public procurement to more restrictive practices. The types are described in Table CS 4.3.

To secure competition, option A, open public procurement, is the nor-mal case for procurement. The use of the more restrictive options B, C,

Table CS 4.3   Types of Public Procurement

Procurement type	Features	Note
A. Open public	There can be unlimited participants.	This is the rule.
B. Limited (not open)	Open invitation exists for expression of interest; short-listed companies are invited to submit proposals.	
C. Negotiated	With or without public invitation to participate, client negotiates with one or several service providers.	It should be used only in rare circumstances.
D. Competitive dialogue	After an EU-wide announcement, the client negotiates with several providers. Using those negotiations, the client elaborates a functional description of the solution, and the participating providers submit their binding proposals. The contract is awarded to the most economic offer.	This procedure was added in 2005 to be used for very complex projects above the EU limits.

Source: Based on DENA (2008a).

and D has to be justified. For example, option C can be used if the service to be rendered cannot be described unequivocally and exhaustively, such that a price for the service can be determined objectively. This applies in EPC projects, especially complex projects such as those covering many federal properties. Since the contractor is expected to determine the specific measures in an EPC project, the objective of the contract is described in a functional way: percentage reduction of energy costs. Option B is also frequently applied in the case of EPCs.

Energy audits. An initial energy audit is usually not required or performed. Baseline technical, energy use, and cost data for the property is provided by the client, based on model data sheets provided in EPC guidelines.[12] Frequently, energy agencies and other external experts support this data gathering. This is usually complemented by facility walkthroughs by bidders. In a two-stage contract a detailed analysis is carried out by the winning bidder, including an audit (see Box CS 4.1). The ESCO will be remunerated for the cost.

To determine the baseline, one needs to gather the following: measured data (at least one year for each meter) for electricity, heat, and sometimes water; profile of facility use; reference prices for utilities; climate

Box CS 4.1

Two-Stage Procedure for Energy Performance Contracts (EPCs)

For complex projects, for example, hospitals or research institutions, a two-stage EPC process is usually applied. In this case, an EPC is concluded with the winning bidder. The contractor then carries out a detailed analysis of the property and verifies the initially proposed energy cost-saving measures and their results and the necessary investments. This step could take two to four months. If the results of the detailed analysis deviate significantly from the first rough analysis, the client can stop the project without any payment to the contractor. Otherwise, the contractor is paid for the analysis and then proceeds with the implementation of the measures.

Source: Hessen Leitfaden 1998/2003; DENA 2008a.

data and correction; and description of the property, its energy facilities, and the building shell, such as windows and insulation. Baseline data are typically collected by the client, sometimes with help from specialists, such as an energy agency or a private consulting company. For the first demonstration projects in a program, the service is often provided free to the client, but later the client has to pay for such services. Many German federal and regional energy agencies have put together guides and data sheet templates for the process.

On the basis of data collected and the energy-saving potential, the objectives of the project will be defined, including the maximum term of the contract; the approximate size of the investment; obligatory measures such as a change in energy source or replacement of windows, and other requirements. The minimum energy cost reduction can then be determined. It will be the main parameter of the EPC.

Types of energy savings projects. The contract typically includes a variety of measures, both equipment (lighting, HVAC, pumps, motors, control systems, etc.) and services (planning; installation, optimization, and maintenance of equipment; controlling; documentation). Frequently, the public client requires some measures, for example, new boilers, and also often thermal renovation measures, such as new windows and insulation, that would not pay back within 10 years. The client may provide funding for these measures, to be integrated into the EPC.

Multiyear contracts. On average, contracts have a term of about 10 years. Durations range from 7 to more than 12 years, depending on the savings potential, the level of investment, and cost saving participation of the client. For contracts of less than 10 years, fewer budgetary issues arise for state and municipal projects. Bigger, longer-term, and federal EP contracts need to be approved and receive a commitment authorization for future payments to the ESCO.

RFPs and proposal evaluations. The RFP contains a functional description of the project: to implement energy saving measures that lead to at least X percent of energy cost savings. It includes all relevant evaluation criteria and weightings, a data sheet with baseline information, and obligatory measures, if applicable. The bids should contain the following information:

- Guaranteed annual energy cost savings (separated by energy source)
- Client share in the guaranteed savings
- Client share in any savings above the guaranteed limit
- Investment cost for equipment and for planning
- Financing concept (including provisions for forfeiting, if applicable)
- Staff training and know-how transfer
- User motivation (if applicable)
- Costs and timing of detailed analysis (if applicable)
- Length of installation period
- Detailed description of energy saving measures and their costs
- Description of automation/monitoring system
- Operations and maintenance plan
- Use of environmentally sustainable technologies

Availability of competent bidders has not been a problem in Germany. When an energy performance contract is announced, 5, 10, and sometimes more ESCOs usually express interest in participating, of which a short list of around three to five bidders is formed, based on the experience and financial capacity of the applicants. At the federal level, with more complex projects with larger investment amounts, the number of participating bidders could be smaller.

The various German procurement guidelines for EPC contain a list of monetary and nonmonetary criteria to evaluate the bids received, including a hypothetical case in which the client carries out the project on its own with its own financing (internal solution; see below). The monetary criteria can be expressed as a net present value, or NPV, calculation. For example, for EPCs in federal properties, DENA (2008a) defines the NPV as the sum of annual net savings to the client, discounted over a period of 15 years. Net savings are computed as follows:

- Energy cost savings
- Plus operating cost savings (as percentage of investment cost)
- Plus value of investments after end of contract term
- Minus contractor's share in energy cost savings
- Minus client contribution to investment (if applicable)

The achievement of nonmonetary, qualitative criteria can be rated on a scale of zero to 100 and thus be combined with the capital value ("Nutzwertanalyse"; see DENA 2008a). The NPV is the dominating criterion, with about 75 percent weight; the other criteria receive about 5 percent weight each (see Table CS 4.4). The valuation can be varied according to the project and the client.

Every bid undergoes this evaluation, including the hypothetical internal solution (see Box CS 4.2). The bidder whose offer receives the highest number of points will be awarded the contract. If the internal solution offers the best value, the tendering process can be canceled.

Project financing. For almost all public sector EPC projects in Germany financing is provided by the bidder. In some cases the client will provide part of the financing, especially for obligatory measures. Although some ESCOs are able to use their own equity for financing, the usual source is

Table CS 4.4   Monetary and Nonmonetary Selection Criteria

Criteria	Weighting	Performance Criteria (0-100 Points)	Weighted Score
Net present value [€]	75%	P1	Q1 = P1*0.75
Bonus participation [%]	5%	P2	Q2 = P2*0.05
Financial proposal and risk mitigation measures	5%	P3	Q3 = P3*0.05
Technical and organizational approach	5%	P4	Q4 = P4*0.05
Quality and compatibility of components	5%	P5	Q5 = P5*0.05
Environmentally friendly technologies, CO_2 reduction	5%	P6	Q6 = P6*0.05
SUM	100%		Sum (Q1 – Q6)

Source: DENA 2008a.

Box CS 4.2

Comparison of Bids with Internal Solutions

EPC guidelines specify how to compare the offers of bidders to a hypothetical internal solution. For an internal solution, expenditures consist of investment costs, including planning costs and construction costs, and operations and maintenance, or O&M, costs. The estimation of energy savings to be achieved through the internal solution must take into account that pressure to achieve guaranteed savings will not be present as in the case of EPC. Therefore, it has to be assumed that energy savings will be lower, typically about 10 to 20 percent.

The specification of the internal solution is problematic. Frequently, the costs of carrying out a project internally are not well known. The guidelines provide rough estimates, such as planning costs in the range of 20 to 25 percent of construction costs.

Source: DENA 2008a.

long-term commercial bank loans. ESCOs can obtain lower-cost loans (about 1 percent below market rates, resulting in interest rates of 3 to 4 percent), comparable to what is available for public sector borrowing, if they sell their future payments from the client to the financing institution (forfeiting or factoring). The forfeiting share is usually restricted to 70 to 80 percent of the agreed payment to the contractor. This is roughly the value of the guaranteed investment, which could serve as a security.

Performance guarantees. The energy performance contract guarantees a minimum percentage energy cost savings over the term of the contract. This limits the payments of the public client to the ESCO. Frequently an immediate sharing of the benefits by the client is agreed upon, say 10 percent. In case the contractor does not achieve the guaranteed cost savings in a period, he has to provide a compensation payment. The public client also has several other ways to reduce the risk of nonperformance. Typically, the installed equipment passes into the ownership of the client after commissioning. The ESCO can also be required (though not in energy performance contract for federal properties) to provide a bank letter of guarantee to the client. Typically, the guarantee covers about 5 percent of the total value of the contractor's share of the guaranteed savings, increasing to 10 percent with forfeiting.

M&V procedures. M&V procedures under energy performance contracts in Germany are fairly straightforward. Energy cost savings can be verified relatively easily through energy bills, metering of energy consumption, and correction of the baseline based on any changes in energy prices, climate, and facility use. Model contracts (e.g., of BEA or DENA) oblige the contractor to install an energy management system (EMS) that includes metering to control the energy consumption in relevant parts of the facilities. Remote monitoring enables the contractor to constantly optimize operation to achieve the targeted savings. It is customary for the contractor and client staff to meet several times a year to discuss any issues, such as changes in the use of the facility. So far, no cases of disputes have been reported.

Project facilitators and energy agencies. In most cases, public sector clients use the services of an external adviser for support in managing the EPC process. An adviser often manages the entire project including the following:

- The selection of properties and recommendations for pooling, if necessary
- The drawing up all papers for contract and award
- Recommendations for technical and commercial targets
- Formal and technical coordination of the tendering process
- The applying of knowledge of the market to facilitate technical and economic appraisal and negotiation of bids
- Recommendation for award of contracts
- Supervision of commissioning of equipment
- Project controlling

Energy agencies, which exist on the federal, state, and communal levels in Germany, often provide those advisory services. In many German states, they have been the first promoters of EPC in the public sector and have developed demonstration projects, guidelines, model contracts, and evaluations of EPC programs. During the first phases of EPC in a state or city, the energy agency typically receives funding from the government and provides its services free of charge. In later stages, funding of energy agencies is shifted to other priorities, and clients are supposed to pay for the services.

After more than 10 years of experience with EPCs in the public sector, project facilitators are still the norm and are needed to spread the word and support public sector agencies that have never carried out such projects. EPC has not yet become a mainstream way of doing business, and many insiders consider the free support to public sector clients by energy agencies essential to keep EPC in the public sector going.

Results and lessons learned. German governments at all levels have been able to successfully establish EPC for public properties. Annually public properties with energy costs of about €20 million are tendering for EPC or ESC. About 2,000 public properties have entered into energy performance contracts or energy supply contracts between 1990 and 2005.

Two recent surveys of local administrations (large, medium, and small cities; see Meyer-Renschhausen 2008) found that 20 percent of medium and small cities have used EPCs. Among cities of more than 100,000, the proportion increased to 32 percent. For ESCs, the respective shares are 36 percent and 56 percent. Interestingly, more users of EPC are satisfied with the results (51 percent) than users of ESCs (31.5 percent). Obviously, barriers exist that prevent more common use of EPC. According to the surveys, 48 percent of local administrations think that they can carry out measures more cost-effectively. Loss of control over properties and uncertainties about budgetary and procurement issues were other reasons cited for not pursuing energy performance contracts, despite their benefits.

It is easier for municipalities to implement EPC successfully when they are knowledgeable partners and know their strength and limitations (Meyer-Renschhausen 2008). Thus cities are generally better able to implement energy performance contracts when they (a) establish long-term strategic planning, (b) create energy data banks, (c) implement continuous energy controlling, (d) bundle responsibilities, (e) communicate success to citizens and make them part of the process, (f) identify high-level champions, and (g) constantly check for support programs.

Practical experience shows that the following factors facilitate energy performance contracting:

- Decision makers who take on the responsibility
- Existence of champions, especially in construction administrations
- A reliable legal framework, with clear information that EPC is allowed, on procurement and contracting procedures
- Standardized procedures and contracts that reduce the time and cost of preparation and increase the reliability of the process
- Neutral process management through a trustworthy facilitator who provides technical and economic know-how

- Formation of pools of properties and buildings for EPC to create economies of scale (Pools should consist of properties within one agency or budget, such as schools or hospitals.)

Among the other, more specific lessons learned about preparing EPC projects are these: (a) forfeiting can lower EPC financing costs; (b) benefits should be shared among all agencies that provide funding for investment and for operating expenditures, by assigning separate shares in the savings in the contract; (c) clearly explaining the client's conditions and expectations in the RFP is important; and (d) consideration should be given to providing public guarantees for small and medium enterprises, to allow them to participate in EPC and thus enlarge the pool of potential bidders.

Even though many of those conditions are met in Germany, mainstreaming EPC has been slow, as we have said, and EPC in the public sector does not seem to be self-sustaining. (see Table CS 4.5) One reason is that because of its federal system, the German practice of public procurement is fragmented. This impedes new ways of doing business both with and within the public sector. Few companies or ESCOs have the resources and specialists to deal with up to 16 different sets of basic tendering documents and procurement procedures. EPC at the municipal level must be authorized by the supervising authority. Differing interpretations of and knowledge about energy performance contracts among those authorities create uncertainty about whether they can be implemented.

Table CS 4.5   Summary of Typical Features of Energy Performance Contracting (EPC) in Germany

Issue	How it's typically done in Germany
Audit	An audit is usually not required or performed. Technical information on the property and energy use and cost data (baseline) are provided by the client, based on model data sheets provided in EPC guidelines. Frequently energy agencies and other external experts support data gathering. This is usually complemented by facility walk-through by bidders. Only in a two-stage contract is a detailed analysis by the winning bidder carried out, including an audit. This is part of the cost of the project, for which the ESCO will be remunerated.
Types of measures	Typically, a variety of measures is included in the contract, both equipment (lighting, HVAC, pumps, motors, control systems, etc.) and services, depending on the type of facility and specific needs; many contracts include obligatory measures (e.g., new boiler, small combined heat and power, windows, or maintenance), which are often financed by the client.

(continued)

Table CS 4.5   Summary of Typical Features of Energy Performance Contracting (EPC) in Germany *(continued)*

Issue	How it's typically done in Germany
Evaluation criteria for ESCO selection	Guaranteed energy savings over lifetime of contract, guaranteed investment, client share of savings above those guaranteed, structure of the offer, project management (investment, O&M, replacement, EMS), CO_2 reduction, etc. Evaluation according to NPV of net energy savings, supplemented with "soft" criteria. Weights vary according to project/client, but the NPV usually has a weight of 75 percent. All offers, plus the equivalent internal solution, must be considered.
Procurement	EPCs are compatible with German/EU procurement rules. Depending on the emphasis, either construction or service contract can be chosen, mostly under limited or negotiated procurement. Model procurement documents and model EP contracts have been developed by several energy agencies for use in German states and by federal authorities.
Multiyear contracts	On average, contracts have a term of about 10 years. For contracts under 10 years fewer budgetary issues arise in the case of state and municipal projects. Bigger, longer-term, and federal EP contracts need to be authorized and receive a commitment authorization for future payments to the ESCO.
Procurement lead time	About one year for project preparation, including procurement; slightly longer for two-stage contracts.
Financing	Usually provided by the contractor with commercial bank loan, in some cases also from its equity. Loan costs are often slightly higher than for municipal loans, but with factoring/forfeiting, similar conditions can be achieved. Client may provide cofinancing if obligatory measures are included.
Performance guarantee, insurance, recourse	ESCO guarantees minimum savings (e.g., 30 percent); sharing of savings is agreed upon in contract; savings usually begin at commissioning, leading to a reduction of public expenditures by perhaps 10 percent. Public client pays about 80 percent of agreed payments to contractor according to the value of its share of the guaranteed savings, in case actual savings are lower; otherwise problems with public budgeting rules arise. Alternatively, the ESCO has to provide a bank letter of guarantee (except for contracting in federal properties), typically 5 percent of the total value of its share of the guaranteed savings, increasing to 10 percent with forfeiting.
M&V procedures	Procedures are comparatively straightforward in Germany. Energy cost savings can be verified relatively easily through energy bills, metering of energy consumption, and adjustments to the baseline. Model contracts oblige the contractor to install EMSs, including meters, to track energy consumption in relevant parts of the facilities. Remote

(continued)

Table CS 4.5   Summary of Typical Features of Energy Performance Contracting (EPC) in Germany (continued)

Issue	How it's typically done in Germany
	monitoring by the contractor seems to be the norm. Very few, if any, disputes have occurred so far.
Incentives, e.g., for client staff	Typically none. Awards exist for best EPC projects. The incentive for EPCs is needed for renovation of facilities and replacement of equipment given the lack of public funds for investments. In many municipalities and states, as well as the federal government, contracting has become essential in climate protection programs.
Support from external experts	Very frequently, especially in the early phase of a program when clients' knowledge about the EPC concept and their trust need to be built up; energy agencies develop guidelines, model contracts, etc. Experts often manage entire project development, preparation, and monitoring process against a fee payment by the public client. After more than 10 years of experience with EPC in the public sector, the concept has not become mainstream. Project facilitators are still the norm and are required to spread the word and support agencies that have not yet carried out such projects.
Bundling/pooling of properties	Depends on level of energy costs and similarity of buildings; advantageous to reduce transaction costs and to include buildings that are less attractive for energy efficiency investment.

Source: Author, based on information cited elsewhere in this case study.

Many insiders in Germany believe that to move the process forward and establish energy performance contracts as a mainstream way of saving energy in the private sector, more extensive use of energy agencies and other experts, providing free advice and project management, is essential.

Acknowledgments

This case study was prepared by Anke Meyer. Valuable information and clarifications in developing the study were provided by the following individuals: Friedrich Seefeldt (Prognos AG), Helga Feidt (Bremer Energie-Konsens GmbH), Michael Geißler (BEA), Michael Hartmann-Skörup (Siemens Building Technologies), Petra Bühner (DENA), Rüdiger Haake (ESCO-Forum im ZVEI), Stefan Scherz (MVV Energiedienstleistungen Berlin), and Werner Neumann (Energiereferat Stadt Frankfurt).

Notes

1. Perlwitz et al. 2005, cited in Geissler and Waldmann 2006. Geissler (2008) estimates that by the end of 2007 more than 100,000 contracts had been signed and about 280 active contractors were generating total annual revenue

of more than €2 billion, in about 170,000 realized projects, representing 12 percent of the total market potential.

2. Based on Bertoldi et al. 2005.

3. The 24 members of the ESCO-Forum (http://www.zvei.org/index.php?id=3708) focus on private sector commercial and industrial clients; the 230 members of the Verband für Wärmewirtschaft (VfW, http://www.energie contracting.de/) are mostly smaller heat delivery service suppliers.

4. The energy saving program in federal properties provides up to €720 million (2006–2011) for energy efficiency measures. Its implementation has been slow. Of €377 million committed to September 2007, for 1,100 proposals, only €8 million was disbursed in 2006, €45 million in 2007, and €32 million in 2008 (as of mid-September). The slow pace is due to necessary planning and permitting processes, but also bureaucratic hurdles (see Deutscher Bundestag 2008).

5. The first-ever EPC in the public sector identified dates from 1991, for the Neue Staatsgalerie Stuttgart in the state of Baden-Wuerttemberg.

6. The federal German Energy Agency (Deutsche Energie-Agentur, or DENA) supports the implementation of the pilot projects.

7. These guidelines specify the various steps of the process of EPC. Almost all guidelines are published in the German language only. The publication *Energypronet 2004* provides an English language overview of EPC that is similar to the guidelines without getting into all the details of procurement and budgetary requirements.

8. Based on the guidelines of the *Hessen Leitfaden*, 1998.

9. For details see DENA 2008b. In the states of Brandenburg, Lower Saxony, and Saxony, energy performance contracts are automatically authorized if profitability can be proved. In Bavaria, Thuringia, and Hessen, contracts under certain amounts do not have to be authorized individually.

10. Local authorities have to introduce a new financial management system by 2010, including adoption of double-entry bookkeeping; see, e.g., http://www .neues-kommunales-finanzmanagement.de/cms/website.php?id=/index.htm.

11. Hospitals have the highest specific energy use of any type of public building. It is estimated that energy savings in German hospitals could result in cost savings of about €600 million and CO_2 reductions of 6 million tons (see Energie sparendes Krankenhaus: Home). However, with more and more municipalities putting their hospitals into private or nonprofit trusteeship, ESCOs are finding it increasingly difficult to secure financing for this sector, which is no longer considered a low-risk public client by the banks.

12. For federal properties, DENA 2008a is relevant.

References

Bertoldi, Paolo, and Sylvia Rezessy. 2005. *Energy Service Companies in Europe: Status Report.* European Commission, Joint Research Centre, EUR 21646 EN.

Brand, M., and Michael Geissler. 2003. "Innovations in CHP and Lighting: Best Practice in the Public and Building Sector." *Proceedings of the First Pan-European Conference on Energy Service Companies,* Milan, May 2003. Ed. Paolo Bertoldi.

DENA. 2008a. *Leitfaden Energiespar-Contracting.* Berlin, September 2008 (first pub. 1999). http://www.zukunft-haus.info/fileadmin/zukunft-haus/publikationen/Leitfaden_Energiespar-Contracting_2008.pdf.

———. 2008b. *Contracting-Lotse für Kommunen. Berlin.* 2008.

Deutscher Bundestag. 2008. Drucksache 16/8676 16. Wahlperiode 31. 03. 2008. Antwort der Bundesregierung auf die Kleine Anfrage der Abgeordneten Peter Hettlich, Winfried Hermann, Dr. Anton Hofreiter, weiterer Abgeordneter und der Fraktion BÜNDNIS 90/ DIE GR.

Geissler, M., Waldmann, A. Goldmann, R. 2006. "Market Development for Energy Services in the European Union." *Proceedings of the ACEEE Summer Study 2006.*

Hessen Leitfaden 1998/2003. Hessisches Ministerium für Umwelt, ländlichen Raum und Verbraucherschutz (1998/2003). Leitfaden für Energiespar-Contracting in öffentlichen Liegenschaften. Bearbeiter: Berliner Energieagentur GmbH, Anwaltskanzlei Schlawien Naab Partnerschaft. Wiesbaden. http://www.hmulv.hessen.de/irj/HMULV_Internet?cid=d779d1 c3557e8dc56b55c3d9c732fdfb.

Kuhn, Vollrad. 2006. *Performance Contracting and Supply Contracting for Several Building Sectors—The German Market and Successful Projects.* Presentation of the Berlin Energy Agency at Energy Days Brussels, March 9. http://www.observatoiredulogemen.

Meyer-Renschhausen, Martin. 2008. Einspar- und Anlagencontracting aus der Sicht des kommunalen Energiemanagements - Ergebnisse zweier bundesweiter Befragungen aus den Jahren 2006 und 2007. http://www.berliner-impulse.de/fileadmin/Berliner_Energietage/2008/.

Perlwitz, H., S. Cypra, D. Möst, and O. Rentz. 2005. "Energie-Contracting, Der Markt in Deutschland." *BWK-Das Energie Fachmagazin,* June 2005, 6.

Prognos AG. 2006a. "Potenziale für Energieeinsparung und Energieeffizienz im Lichte aktueller Preisentwicklungen." Report 18/06 for the German Ministry of Economy and Technology, Berlin. http://www.bmwi.de/BMWi/Redaktion/PDF/Publikationen/Studien/studie-prognos-energie.

———. 2006b. "Contracting-Potenzial in Oeffentlichen Liegenschaften." Report for DENA. See summary: 2049_Marktstudie Contracting.

Seefeldt, Friedrich. 2003. "Energy Performance Contracting: Success in Austria and Germany—Dead End for Europe?" *Proceedings of the ECEEE Summer Study 2003.* http://www.eceee.org/conference_proceedings/eceee/2003c/Panel_5/5158seefeldt/.

World Engineers Convention 2008. "Engineering: Innovation with Social Responsibility" Brasilia, December 3–5, 2008. http://www.wec2008.org.br.

Energy Performance Contracting in Japan

The energy service company, or ESCO, business does not have a long history in Japan. It was introduced in 1996 when the Resources and Energy Agency delegated a special committee to learn about ESCOs. From 1997 to 1998, extensive research was conducted by the Energy Conservation Center, Japan (ECCJ), a public organization designated to promote the efficient use of energy, climate change mitigation, and sustainable development, including the exploration of new business models and energy policies. In 1999, 16 companies created the Japan Association of Energy Service Companies (JAESCO); today membership has increased to 132.

The ESCO market in Japan had grown to 40.6 billion yen (US$406 million) by fiscal 2007, a 46 percent increase over fiscal 2006, and the number of energy performance contracts reached 176. The number of public sector ESCO projects in fiscal 2006 was 50. Energy intensity improvement has averaged 12 percent per project. Reports indicate that CO_2 emissions reduction as a result of the ESCO business is at least 1.09 million tons of CO_2 per year, about 15 to 17 percent of the total mitigation expected from the plan to achieve Kyoto protocol targets. The total potential ESCO market is said to be 2,470 billion yen, and the potential market in the public sector is estimated to be about 600 billion yen.

The major ESCO business models in the public sector are guaranteed savings and shared savings. Under the guaranteed savings scheme, the initial investment for renovations is paid by municipalities. The ESCO guarantees the energy cost savings, which generate a revenue stream of which part will repay the investment costs and part will be paid to the ESCO. Under a shared savings contract, the ESCO is responsible for financing. In this case, municipalities pay a greater portion of the energy savings to the ESCO to cover its fees and financing costs. Currently, the public sector market is shifting toward shared savings contracts, which now account for almost 70 percent of projects. The main reason for the shift is the current weak financial condition of many municipalities in Japan.

Subsidies play a big role in the Japanese ESCO business, and currently 20 to 30 different ones are available (see Table CS 5.1). These are provided through organizations such as the New Energy and Industrial Technology Development Organization (NEDO); the Ministry of Environment; and the Ministry of Land, Infrastructure, Transport and Tourism. The subsidy is typically about one-third of the total cost of the project.

There is no standard ESCO program in Japan. However, to promote the ESCO business ECCJ has developed optional guidelines for use by municipalities, which are summarized in this case study.

An energy audit. The energy audit consists of a preliminary energy diagnosis, a walk-through audit, and a detailed feasibility study. The preliminary diagnosis can be carried out in three ways:

- A free energy diagnosis can be performed by ECCJ; it focuses on the operational management of the building and does not necessarily involve the ESCO business;
- The agency can pay a third party, such as a consulting firm, to conduct a diagnosis; or
- It can apply for a grant from NEDO to hire an external firm to conduct an audit. This last option is available only to municipalities that agree to use the ESCO approach.

According to the diagnosis, the public sector entity develops a Request for Proposal (RFP) that provides the following data: an outline of the facility; complete drawings, including architectural designs, and description of the power, sanitation, and HVAC systems; current capacity utilization; and specifics of central monitoring, facility maintenance, automatic control maintenance, and so forth.

Table CS 5.1 Examples of Major Subsidies

Subsidies	Target	Subsidy amount	Approvals/ Applicants	ESCO
NEDO[a] Support to business for efficient use of energy.	Existing factory or office with high expected impact.	Up to 1/3 of cost, to max 500M yen (total 15.9B yen).	140 / 159 (88%)	Yes
NEDO Promote installation of high-efficiency energy systems (buildings).	Industrial buildings (15% savings for new, 25% savings for existing).	Unlimited up to 1/3 of cost (total 1.5B yen).	31 / 34 (91%)	Yes
NEDO Promote installation of high-efficiency building energy mgmt. systems (BEMS).	BEMS in new and existing buildings; more than 1% savings required.	Up to 1/3 of cost, to max 100M yen (total 2.7B yen).	59 / 60 (98%)	Yes
Japan Electro-Heat Center Promote installation of high-efficiency energy systems (e.g., air-conditioning).	Need to introduce high-efficiency air-conditioning.	1/3 higher efficiency than existing air-conditioner.	76 / ?	Yes
Heat Pump and Thermal Storage Technology Center of Japan (HPTCJ) Disseminate load utilization systems.	Facilities with high load utilization, such as NaS battery and cold storage.	Up to 1/3 of cost; max unknown (total 1B yen).	16 / ?	Yes
NEDO Comprehensive energy saving promotion led by energy supply companies.	Coapplication by energy supply companies, local public agency, and owner required; introduce energy saving system in more than 2 consumer buildings to save energy and water.	Up to 1/2 of cost; max unknown (total 1.7B yen).	6 / 6 (100%)	Yes
Ministry of Environment Subsidies for local public agencies.	Introduce energy saving facility;[b] require more than 10% CO_2 reduction at a cost less than 10,000 yen/ton; not applicable for shared savings.	Up to 1/2 of cost; max unknown (total 1.64B yen).	17 / ?	Only municipal ESCO projects.

Source: Japan Facility Solutions, Inc. (JFS) 2008.

a. NEDO = New Energy and Industrial Technology Development Organization.

b. Based on action plan determined by the Law Concerning the Promotion of Measures to Cope with Global Warming.

c. ? = Number of applicants not known.

After the RFP is developed, bidders are allowed to conduct walk-through surveys to collect additional data that they need to prepare a proposal. This opportunity is usually offered once or twice, depending on the needs of the bidding ESCOs. Finally, after an ESCO is selected, it will conduct a detailed feasibility study. The whole process from initial diagnosis to the feasibility study can take up to two years (see Table CS 5.2).

Evaluation criteria. Once a municipality decides to develop an ESCO project, it must appoint a selection board of mostly internal members. The selection committee will discuss the evaluation system and criteria and announce the evaluation process officially. Criteria typically cover the bidders' project financing plans, technical proposals, their plans for managing operations and maintenance (O&M) and measurement and verification (M&V), and project management plans. Table CS 5.3 includes sample criteria for municipality-led projects. It represents a standard selection process, but municipalities can customize it by weighting the criteria based on their preferences and including additional ones if desired.

The public announcement of the evaluation results should be decided upfront and clarified in the RFP. Each score is disclosed to the individual bidder, so that the selection committee can address questions from bidders after the selection process has been completed. The highest-ranked ESCOs, based on the written proposals, are invited to make an oral presentation to the agency's selection committee on their approaches to the project. A combination of the proposal score and the oral interview will ultimately determine the ESCO selected for a project.

Table CS 5.2   The Energy Audit Process

Energy audit	Level	When to conduct	Days to complete
Saving energy diagnosis	Preliminary: Minimum technical data and analysis, energy demand and use profiles, indication of potential	Basic data gathering to identify buildings that may or may not have energy savings potential	—
Walk-through	Preliminary: Preliminary savings estimates	Prior to bidding on RFP; conducted and paid for by the ESCO	1 to 2
Feasibility study	Detail: Detailed saving estimates to ensure effectiveness	After ESCO selection; conducted and paid for by the ESCO	About 65

Source: ECCJ, 2008.

Table CS 5.3 Selection Criteria for the Qualified Bidders List

Criteria	Points available	Example 1 Weighting factor	Total	Example 2 Weighting factor	Total
Attractive benefit within 15 years.	5	5	25	4	20
Attractive annual benefit to the municipality within the contract period is proposed.	5	5	25	4	20
Substantial guaranteed savings in utilities is projected.	5	5	25	4	20
The financing plan is reliable.	5	4	20	5	25
The contract period is as short as possible.	5	3	15	3	15
The proposal related to possible subsidies is included.	5	2	10	3	15
The project includes sufficient energy savings.	5	5	25	5	25
Global warming countermeasures are considered, such as measurable reduction in CO_2 emissions.	5	5	25	3	15
Environmental impact is considered, such as NO_x, SO_x, dust, and noise.	5	2	10	2	10
The technology and proposal are specific and reasonable.	5	4	20	5	25
The proposal is unique and indicates particular know-how.	5	2	10	4	20
Renovation rather than the renewal of existing facilities is considered.	5	2	10	3	15
Methods proposed for maintenance, monitoring, and operation control are specific and reasonable.	5	4	20	5	25
The potential bidder has the ability to provide excellent product quality control, complete construction on time, and provide service.	5	2	10	4	20
A proposal is included for support after the contract period is finished.	5	1	5	2	10
Balance and excellence of the total proposal.	5	5	25	5	25

Source: Energy Conservation Center, Japan.

Note: Example 1 is from an ESCO project at a rehabilitation center; Example 2 is from a project at a municipal building. The total number of points available can vary depending on the project.

Multiyear Energy Performance Contracts (EPCs). ESCOs can propose a contract period not to exceed 15 years. It usually takes seven to eight years to recover project investments. Guaranteed savings contracts used to be limited to five years because it was difficult for municipalities to take on longer-term debt. However, a new law on environmentally conscious or green contracting, which went into effect in November 2008, allows municipalities to take on debt for up to 10 years.

Project financing. Under the RFP, the public agency will define the type of contract to be used (i.e., guaranteed or shared savings). In guaranteed savings contracts, the total project cost will also be defined in the RFP, so that the ESCO can prepare its proposal according to the specified budget. In shared savings contracts, the source and amount of project financing are not predetermined and may be specified by the ESCO in its proposal. Because most Japanese ESCOs are relatively small, they cannot mobilize large-scale financing on their own, and thus most projects have used leasing companies to assist with financing. Another advantage of using leasing companies is that neither the public agency nor the ESCO has had to include the project debt on its balance sheet. However, a reform in the accounting system in April 2008 disallowed such practices, and now either the public agency or the ESCO is required to show these investment liabilities on its books. This new requirement may, unfortunately, undermine the government's desire to further promote the ESCO business. Experts think the law will result in a reduction in registered ESCOs in Japan and may limit them in the future to large energy companies or subsidiaries of major manufacturing conglomerates.

Performance guarantees. Payments to the ESCOs are based solely on the actual energy and water savings achieved from the energy efficiency project. The contract includes an energy savings performance guarantee over the life of the contract. If there is a shortfall in the actual energy savings, the value of the shortfall is deducted from the energy service fee. If project performance exceeds the guarantee, the additional benefits will be shared between the ESCO and public agency according to the contract.

M&V procedures. The frequency of monitoring reports from the ESCO depends on the stage of the project. There is no standard or mandatory process. Generally ESCOs develop a report twice a year, but in the first year it is common to monitor project performance more frequently. The ESCO proposes an M&V plan, which must be approved by the municipality. This gives the ESCO flexibility to develop the most suitable M&V plan, based on the project and the specific energy efficiency measures used.

In the event that the public agency has concerns over the quality or credibility of the report, it is allowed to hire a third party to conduct independent M&V. If the results from these measurements contradict the ESCO reports, the agency has a right to recover its costs and the ESCO must revise its M&V methodology and seek approval from the public client.

Incentives for public sector staff. Public staff members associated with ESCO projects do not receive any direct financial incentives or commissions. The increase in the ESCO business is driven in part by recent energy price increases and by tighter regulations for business facilities under the Energy Saving Act amendment. Japan also is committed to achieving its emissions reduction target under the Kyoto Protocol, which, according to government calculations, requires additional reductions of 20 million to 34 million tons of CO_2 per year. The government hopes that much of this additional required reduction can be achieved under the new Energy Savings Act amendment.

Acknowledgments

This case study was prepared by Suiko Yoshijima. Valuable information and clarifications in developing the case study were provided by the following individuals: Mr. Furukawa and Mr. Nagumo (General Affairs Department contract and property unit, Itabashi Ward Office), Mr. Keiichi Oguma and Mr. Takahiro Abiko (Japan Facility Solutions), and Mr. Nobuo Tanaka (ECCJ).

References

Energy Conservation Center, Japan, "ESCO donyu no tebiki" ("Guidance for Municipalities on ESCO Introduction") presentation documents, Tokyo, 2008.

JAESCO, *JAESCO Newsletter,* September 2008.

JAESCO Web site, http://www.jaesco.gr.jp/index.htm.

Japan Facility Solutions, Inc. (JFS), "The Proposal on Energy Savings and CO_2 Mitigation Using ESCO Services," presentation documents, Tokyo, 2008.

Research Institute of Economy, Trade and Industry, http://www.rieti.go.jp/jp/events/bbl/05111401.html.

Energy Performance Contracting in the Public Sector in India

India's economy has experienced remarkable growth since the economic liberalization of the early 1990s, and it is likely to continue to be one of the fastest growing in the world. To support the projected growth rate of over 7 percent per annum, India needs to develop additional energy resources. Investments in power generation, transmission, and distribution have not been able to keep up with economic growth. As a result, India's electricity sector currently faces problems of inadequate capacity, poor quality, and unreliability. Current electricity shortages are estimated to be 20 percent for capacity and 15 percent for energy. These shortages are especially detrimental to industry and commerce, which have been the main engines powering India's growth and development.

The inefficiencies in India's existing energy infrastructure and the technical potential for energy efficiency improvement have long been known and are very large. Recognizing the importance of energy efficiency, the Government of India (GOI) enacted the Energy Conservation Act of 2001. Under this act, the Bureau of Energy Efficiency (BEE) has been established as the nodal agency responsible for the improvement of energy efficiency. BEE's mission is to "institutionalise energy efficiency services, promote energy efficiency delivery mechanisms, and provide leadership to improvement of energy efficiency in all sectors of the economy."

Evolution of ESCOs in India

Studies of energy efficiency in India had pointed out the large potential for energy savings, particularly in the commercial, institutional, and industrial sectors. Actual implementation of energy efficiency projects has fallen far short of the potential because of a number of barriers. In 1994, the U.S. Agency for International Development (USAID) sponsored an assessment of the potential applicability of energy saving performance contracting (ESPC) approaches in India and the role of energy service companies (ESCOs) in energy efficiency project implementation. The first ESCO in India was established soon after the publication of this report and developed some private sector ESPC projects.[1]

In 2001, USAID initiated the first stage of the Energy Conservation Commercialization Project (ECO), which studied ways to encourage and facilitate ESCO development and ESPC implementation. Under ECO, case studies of ESCOs and standard performance contracts were developed, and a loan fund was created (managed by ICICI, a local commercial bank) to finance demonstration projects. The first public sector ESPC project financed and implemented under ECO was a street lighting energy efficiency project for the City of Nashik, in Maharashtra, in 2003.[2]

Also in 2003, in cooperation with the Asian Development Bank (ADB), BEE initiated the Energy Efficiency Enhancement Project (EEEP) to develop innovative approaches to foster the development of a sustainable energy efficiency market in India. One of the important elements of this project was the development of financial mechanisms to encourage ESCOs to participate in energy savings projects in the public sector.

As a part of the EEEP, the project team developed an approach for applying the ESPC model in the public sector. This approach included the definition of a financing model involving an ESCO, a financing institution, equipment suppliers, and the host public agency, as well as a "payment security mechanism" that would provide some assurance to the ESCO and the lender that they would receive payment under the ESPC.[3] To demonstrate the application of this approach in the public sector in India, BEE proposed a program to implement energy efficiency measures in high-profile central government buildings.

The Central Public Works Department ESPC. Central government buildings in India are managed by the Central Public Works Department (CPWD); therefore, performance contracting projects are under CPWD purview. BEE provided information and education to CPWD staff members on the ESPC model and on the ESPC procurement and implementation process. CPWD agreed to try the model and in 2003, issued Request for Proposals (RFPs) for a number of central government buildings, using a competitive bidding process:[4]

1. BEE prepared a draft RFP in late 2003 and provided it to CPWD.
2. BEE engaged consultants to conduct energy audits of the target buildings.
3. CPWD refined the RFP and issued it publicly to interested firms in January 2004.
4. CPWD did not include a screening process and thus did not issue a request for expressions of interest (EOI).
5. Proposals were received in May 2004, and the evaluation process was completed in October 2004.
6. Several contractors were selected (after protracted negotiations) for the different facilities, and the contracts were signed in 2004–2005.

Municipal ESPC projects. Municipalities in India (known as "urban local bodies," or ULBs in smaller cities) are attempting to improve the reliability and quality of infrastructure services, including water service delivery and wastewater management. In many cases, ULBs spend more than 50 percent of their operating budgets for energy, and reducing these costs through energy efficiency measures can contribute significantly to their financial viability. A number of studies and energy audits in India have pointed out the large potential for improvement of energy efficiency and reduction of energy costs in ULBs. The most important energy efficiency measures include optimum pumping systems design, upgrading or replacement of street lighting, and improved controls and operational procedures. However, the implementation of such measures by ULBs was constrained by the limited technical knowledge and capability of water utility staff members regarding energy efficiency options and the lack of available capital for financing such projects. Therefore, the ESPC approach was considered a viable mechanism for implementation of energy efficiency projects in these settings.

The Alliance to Save Energy initiated work under the USAID-funded "Watergy" program toward the implementation of ESPC projects for municipalities in Karnataka and other states. This effort identified needed policy reforms, conducted energy audits, developed various tools and resources, and conducted training and capacity building for the ULB staff. Actual implementation of energy efficiency projects was delayed by a number of institutional barriers. In 2005, the World Bank funded a project to develop an ESPC framework for municipal water utilities. Later, BEE published a manual for energy savings projects in municipalities, which focused on the ESPC approach.

An important municipal project was sponsored by the Tamil Nadu Urban Development Fund (TNUDF). This fund was established as an autonomous financial intermediary in 1996 to improve the operational efficiency of Indian municipalities and help them access private capital. TNUDF adapted the ESPC framework and approach defined by the World Bank and in 2007 initiated a project to implement an ESPC in which projects in seven municipalities were bundled.

The Tamil Nadu model has now been replicated in the State of Gujarat by the Gujarat Urban Development Corporation (GUDC), which is a part of the Urban Development Department of the Government of Gujarat, constituted to support municipal infrastructure projects in the state. GUDC has initiated a major ESPC initiative to promote municipal sector energy efficiency projects at seven municipal corporations and 159 urban local bodies in Gujarat. Recently, a new public company has been proposed, Energy Efficiency Services Ltd. Jointly owned by four state companies, including NTPC and PowerGrid Corporation, this new company would help develop the energy efficiency market in all sectors through a range of mechanisms, which may include acting as a public ESCO.

A number of municipalities have implemented street lighting projects using the ESPC process.[5] These projects have involved replacement of existing mercury vapor lamps with efficient fluorescent tube lamps (T-5 lamps) or installation of controls on the lighting circuits.

Other public sector ESPCs. The Indian Renewable Energy Development Authority (IREDA) initiated an ESPC project in public buildings in three States (West Bengal, Gujarat, and Andhra Pradesh). Under this project, RFPs were developed for two buildings in each of the states, and bids were invited from energy service providers (ESPs). However, very few ESPs submitted bids, and only one project was actually contracted. Public building ESPC projects have also been implemented by DSCL ESCO in the All-India Institute of Medical Sciences (Delhi) and the Delhi Municipal Corporation.

Public ESPC Procurement

On the basis of experience with these projects, the art and science of ESPC have achieved a degree of maturity; the newer projects are using standardized RFP and contract formats. The major steps in the ESPC process currently used in the public sector in India are as follows: short-listing of ESCOs using the EOI process; issuing the RFP; evaluating and selecting the ESCO; negotiating the contract; and conducting measurement and verification (M&V).

Short-listing of ESCOs. In the initial stage of the procurement process, the public agency requests expressions of interest from ESCOs. The EOIs are evaluated based on criteria that generally include the following:

- Technical capabilities
  - Identification, assessment, and implementation of energy efficiency measures
  - Management of ESPC projects
  - Monitoring of energy savings
  - Operation and maintenance
- Capabilities in investment grade audits, including availability of equipment for energy auditing
- Experience with ESPC
- Ability to provide guarantees
- Annual turnover

After an evaluation of the responses to the EOIs, ESCOs (generally about 6 to 10) are short-listed for the RFP.[6]

The RFP. The next step is the issuance of the RFP to the short-listed providers. In the case of TNUDF, the RFP was titled "Technical Consulting Services to Conduct Investment Grade Energy Audit (IGA) Study and Implement Energy Efficiency Projects under Performance Contract" and covered seven municipalities in Tamil Nadu. The RFP addressed municipal pumping and street lighting energy efficiency improvement and covered the following services:

- Conduct of an investment grade energy audit (IGA)
- Design and implementation of energy efficiency measures
- Project management and commissioning
- Operations and maintenance of existing and new equipment

Generally the RFPs require a two-phase approach:

Phase 1 — Investment Grade Audit (IGA). In this phase, the ESCO is required to complete the IGA and identify the energy efficiency measures to be included in the investment package. Detailed specification of the requirements for the IGA is usually provided in the RFP. The cost of the IGA is a consideration in the resulting contract agreement. This cost is included in the Phase 2 project costs and reimbursed from the energy savings. In case, for any reason, the public agency does not approve the measures that the ESCO suggests in the IGA, and elects to not proceed to a contract, the agency reimburses the ESCO for the cost of the IGA, and there is no commitment to continue the project with that provider. In most cases, the reimbursable cost of the IGA is limited to a maximum amount.[7] The audits and audit reports then become the property of agency.

Phase 2 — Implementation. This phase includes the ESPC arrangement with the ESCO and completion of the engineering, procurement, construction and commissioning phases, including performance verification. Measurement and verification (M&V) of the energy savings from the installed projects are usually conducted by a mutually agreed-upon, independent third party, which may be proposed either by the public agency or by the ESCO. In some cases, the ESCO conducts the M&V. Regardless of who conducts the M&V, the cost is paid out of the savings generated.

A unique feature of recent RFPs for ESPCs was that the ESCO was required to achieve minimum savings (30 percent in Tamil Nadu, and 20 percent in Gujarat) for both municipal pumping and street lighting. If the IGA indicated lesser savings, not only was the project not implemented, but the IGA costs were not reimbursed. The ESCOs were encouraged to contact and visit (at their own cost) the ULBs before submitting their proposals to ensure that the savings were achievable. The rationale for the minimum savings requirement appears to be to discourage cream-skimming and to ensure significant agency benefits.

The strict requirement for minimum savings of 30 percent in Tamil Nadu may have discouraged some of the short-listed ESCOs from bidding on the project.[8] Those that did submit a bid had a very significant incentive to come up with at least 30 percent savings. The winning bidder, Asian Electronics Limited, committed to the minimum 30 percent savings target but appears to have encountered problems meeting it.[9]

Evaluation criteria. The proposal evaluation process usually consists of three stages. In the first stage, the technical proposals are evaluated using a specified set of criteria. An example is shown in Table CS 6.1. A minimum

Table CS 6.1   Example of Technical Proposal Evaluation Criteria in the Tamil Nadu Urban Development Fund (TNUDF) Request for Proposal (RFP)

Criterion	Points
Specific experience of the consultant firm related to the assignment:	
(a) Experience in a similar assignment to investment grade energy studies in municipalities or municipal corporations	10
(b) Experience in a similar assignment in street lighting and/or water pumping in municipalities or municipal corporations	10
Methodology for the project	50
Key professionals:	
Team leader	
Project manager	
Project engineer	30
Total	100
Minimum Qualifying Score	80

Source: Tamil Nadu Urban Infrastructure Financial Services Ltd 2007.

technical score is also specified. The technical evaluation is conducted before the financial proposals are opened.

After the technical evaluation is completed, financial bids are opened (second stage) only for the bidders meeting or exceeding the minimum technical scores. The financial proposals require the ESCO to designate the cost of the IGA and the percentage of energy savings it can provide to the public agency. The financial evaluation is based primarily on the proposed sharing of the energy savings. The ESCO sharing the largest percentage of the savings with the public agency is selected, provided it met the minimum technical score. The third stage of the process is the contract negotiation.

Performance guarantees. The ESPC process requires the ESCO to provide a financial performance guarantee in the form of a fixed deposit receipt of any scheduled bank in favor of the public agency. This performance guarantee has to be deposited before the award of the contract and be valid for the period of performance of the ESPC.

The ESPC specifies that the performance guarantee is required for full and faithful performance by the ESCO. In case the ESCO fails to perform its obligations under the contract, fails to maintain the progress of the work, or does not complete the work as required under the contract, the public agency reserves the right to acquire the performance guarantee in full.

Trust and retention account. To assure the ESCO that it will be paid for its services in accordance with the shared savings arrangement as proposed and contracted for, an escrow account—generally referred to as a "trust and retention account" (TRA)—is established by the agreement of the public agency and the ESCO. A trustee bank is selected by mutual agreement and the TRA is established at the bank. The bank is instructed that this account is only to be used for releasing payments to the energy efficiency project. The agency escrows the amount of its usual electricity bill (the amount before commencement of the energy efficiency project) into the TRA. The trustee then makes payments in the following order of priority: (1) payment to Electric Utility, (2) payment to the ESCO, (3) payment to the agency, and (4) transfer to reserve fund (if any amount is remaining after making the first three payments).

Conducting M&V. As indicated above, measurement and verification are usually carried out by an independent third party. The ESCO is responsible for providing the measuring and monitoring equipment and specifying the M&V protocol and methodology. Some ESPCs specify M&V monthly for the first three months and then periodically as mutually agreed by the agency and the ESCO.

O&M and training. The ESPC generally requires the ESCO to be responsible for O&M of the energy efficiency measures and the training of appropriate agency personnel as required for properly maintaining the equipment to achieve the savings. An O&M plan is prepared and submitted by the ESCO as a part of the response to the RFP.

Key Issues and Challenges
The past five to six years have seen some high-profile ESPC projects in the public sector in India, particularly the CPWD projects for central government buildings and some recent activity with respect to municipal water pumping and street lighting. These projects have substantially advanced the state of the art of ESPC in India. However, a number of issues and challenges remain.

Budgeting. No explicit provisions in central or state government budgets cover energy savings performance contracting. In India, expenditures are made from two budget types—revenue and capital expenditures. All operations and maintenance expenses fall within the revenue expenditure stream, with budgets allotted based on the previous year's expenditure.

The CPWD has responsibility for both the revenue and capital expenditure budgets for buildings of the central government. The CPWD ESPC projects for central government buildings were funded through the revenue budget under the "planned" expenditure stream.

The situation is more complex at the state level. For example, in Karnataka, payments to utilities for electricity used by the municipality are made by the Department of Finance; therefore, the municipality does not benefit from the savings from an ESPC project. A similar situation occurs in Gujarat with respect to ESPC procurement. Karnataka addressed the issue through Government Order UDD 14 SFC in 2006 that directed 45 urban local bodies in the state to implement performance-based contracting. Such a specific directive supports the ULBs to implement the contracts and pay the shared savings components through their revenue budgets at the state level. A similar government order is being developed in the State of Gujarat by the Finance Department in cooperation with the Public Works Department and the Gujarat Urban Development Company.

Payments from public agencies. The response of private sector ESCOs to ESPC procurements from the public sector has been less than overwhelming because of considerable skepticism on their part about receiving payments from public agencies, particularly at the state level. This was an important factor in the very low response to the state agency RFPs in the IREDA project described previously. The recent projects in Tamil Nadu and Gujarat have addressed this issue by using the trust and retention account approach. The success of these projects will help overcome this problem.

Financing. The projects that have been implemented have used funds from donors such as USAID or municipal infrastructure funds (as was done in Tamil Nadu and Gujarat). Commercial financial institutions have been reluctant to finance public sector projects and have participated only when provided funding from donors. The national Energy Conservation Act requires that each state establish a state energy conservation fund (SECF) to help finance energy efficiency projects. The first such fund is being set up in the state of Kerala in 2009. It has provisions for technical assistance and financing of public sector ESPC projects. As SECFs are established in other states in India (the state of Madhya Pradesh is also in the process of establishing such a fund), additional financing is likely to be available for ESPC projects in the public sector.

Standard RFPs and contracts. A review of recent RFPs and contract templates shows that a certain amount of standardization has already taken place (starting with the initial RFP and contract developed by the CPWD). Such standardization will reduce the time and effort required on the part of both public agencies and ESCOs to implement the ESPC process in future procurements. It should be noted, however, that it has taken a number of years to get to this point, and further changes in the "standard" documents are likely to occur as additional experience with ESPC procurement is gained.

Addressing the Key Issues and Challenges—The Gujarat Project

The most recent municipal ESPC project, currently under way in Gujarat, covers seven municipal corporations and 159 ULBs. The project is being implemented for 10 ULBs in Phase 1 and for all the others in Phase 2. The Gujarat Urban Development Corporation (GUDC) has developed a unique funding arrangement with the Power Trading Corporation (PTC; a government of India undertaking) which will extend the required credit to the ESPC projects. Project management services are being provided by the Infrastructure Leasing and Financing Services (IL&FS) ECOSMART. GUDC, through the Gujarat state government, has advised the ULBs and corporations to create escrow accounts using revenue streams from the ULB projects, which will be used to repay the principal, interest, and utility bills.

GUDC has constituted a project management cell of technical experts from utilities, ULBs, IL&FS, and a USAID ECO III team. The project management cell will support ESP selection, the M&V process, and the payment mechanism. Table CS 6.2 illustrates some of the key features of this project.

Table CS 6.2   Key Features of the Gujarat Municipal ESPC Project

Issue	How addressed
Multiyear contracts	Multiyear contracts are feasible under the GUDC structure.
Retention of savings	ULBs/municipal corporations will retain 20 percent of savings achieved.
Initial energy audits	Funded by the GUDC for 7 municipal corporations and 159 ULBs, with a budget of Rs 7 "crores," or Rs 70 million (US$1.5 million).
Project definition	Project defined by the results of initial audits; covers water and sewage pumping systems and street lighting.
RFP	Standardized format created using modifications and enhancements to the earlier municipal models.

(continued)

Table CS 6.2   Key Features of the Gujarat Municipal ESPC Project *(continued)*

Issue	How addressed
Evaluation criteria	Technical evaluation based on ESCO experience, project management approach, and quality of technical proposal; financial evaluation based on percentage savings to ULBs, minimum investment by ESCO, and financial capacity of ESCO.
Evaluation capacity	Project management cell (IL&FS ECOSMART) established.
Financing source	Financing arranged through the Power Trading Corporation.
Payment to ESPs	Escrow accounts established to ensure payment.
M&V implementation	M&V conducted by third-party contractor.
Contracting capacity	Provided by project management cell.

Source: Authors.

Acknowledgments

This case study was prepared by Dilip Limaye. Valuable information and clarification were provided by the following individuals: Dr. Mahesh Patankar (independent consultant), Mr. Lekhan Thakkar (GUDC), Dr. Nitin Pandit (IIEC), Dr. G. C. Datta Roy (DSCL ESCO), Mr. Madhav Dandavte (Asian Electronics Ltd.), and Mr. R. Vasu (INTESCO Asia Ltd.).

Notes

1. The first ESCO in India is INTESCO-Asia Limited, established in Bangalore.

2. This project was conducted by Sahastronic Controls Private Limited, using financing provided by ICICI Bank under a line of credit from USAID.

3. A similar approach was taken earlier in the financing of an energy efficiency project at Orissa Sponge Iron Corporation, under an initiative by the Indian Renewable Energy Development Agency (IREDA).

4. The buildings included the president's home and office (Rashtrapati Bhawan), the prime minister's office, the Ministry of Power building (Shram Shakti Bhawan), and the Ministry of Railways building.

5. Examples include projects in Akola, Latur, Ujjain, Indore, and Pune Municipal Corporations by Asian Electronics; in Bangalore by ElPro; and in Sangl-Miraj by Sahastronic Controls.

6. For example in the TNUDF ESPC procurement for energy efficiency measures in municipal pumping and street lighting, the short-listed ESPs were Intesco Asia Pvt. Ltd,. DSCL Energy Services Ltd., Agni Energy Services, Dynaspede Integrated Systems, Asian Electronics, Elpro Energy Dimension Pvt .Ltd., Servomax India Ltd., and ES Electronics India Pvt. Ltd.

7. A typical maximum reimbursable IGA cost is Rs 50,000 (about US$1,000).

8. R. Vasu, of INTESCO Asia Ltd., indicated that his firm had doubts about achieving 30 percent savings in municipal pumping and therefore declined to submit a bid (pers. comm., 2008).

9. Asian Electronics implementation team participant, pers. comm., 2008.

References

Bureau of Energy Efficiency, "Action Plan," 2003.

Bureau of Energy Efficiency, *Manual for the Development of Municipal Energy Efficiency Projects*, 2008.

Business Standard, "Four Power PSUs to Establish Energy Efficiency Company," June 23, 2009.

Charles River Associates (Asia-Pacific) Pty. Ltd., Energy Efficiency Enhancement Project (EEEP), "Final Report," 2005.

ECO-Asia Clean Development and Climate Program, *Establishment of the Kerala State Energy Conservation Fund*, report prepared for the Energy Management Centre, Government of Kerala, November 2008.

Government of India, Energy Conservation Act, 2001, *Gazette of India*, September 29, 2001.

Lekhan Thakkar, "Institutional Framework for Municipal DSM—The Gujarat Experience," Gujarat Urban Development Company Limited, 2009.

SRC International, "Feasibility Study on the Introduction of Energy Service Companies in India," January 1995.

Tamil Nadu Urban Infrastructure Financial Services Ltd., "Request for Proposals — Implementation of Municipal Energy Efficiency Projects under Performance Contract, Water Supply and Street Lighting," India, September 2007.

The World Bank, Public-Private Infrastructure Advisory Facility, "A Strategic Framework for the Implementation of Energy Efficiency Projects for India Water Utilities," prepared by SRC Global Inc., September 2005.

USAID, "Evaluation of the Watergy Program in India, Final Report," prepared by Nexant, Inc., September 2005.

World Bank Group Projects with Public Sector Energy Efficiency Components (2000–09)

Project	Year approved	Public sector targeted	Specific components	Use of ESPCs
World Bank				
China—Beijing second environment	2000	Public buildings	Conversion of coal to gas boilers in public and other facilities; some heat energy conservation demonstration projects in public facilities	No
Ukraine—Kiev public buildings energy efficiency	2000	Public buildings	Energy efficiency (EE) improvements in institutional buildings, including health care, educational, and cultural facilities	No
Russia—Municipal water and wastewater	2000	Municipal water supply	Rehabilitation of pumping station, leak reduction, water demand management, EE in wastewater treatment plants	No
Ukraine—Lviv water and wastewater	2001	Municipal water supply	Rehabilitation of pumping stations and replacement of pumping units in wastewater facilities	No
Poland—Krakow energy efficiency	2001	Municipal buildings and housing blocks	Creation of utility-based ESCO (POE ESCO) within district heating (DH) utility (MPEC) to provide services to public buildings for improved weatherization, on-site boilers, electrical equipment (motors, pumps, lighting)	Yes
Belarus—Social infrastructure retrofitting	2001	Schools and hospitals	Energy retrofitting of schools and medical facilities (building envelope and heating system improvements, conversion or replacement of individual boilers)	No
Ecuador—Power and communications sectors modernization and rural services	2001	Street lighting	Market development efforts for EE that include options to finance and implement improvements in public lighting	No

Project	Year	Sector	Description	
Lithuania—Education improvement	2002	Schools	The upgrade of school facilities to improve heating and ventilation systems and building envelope measures	No
Croatia—Energy efficiency	2003	Schools, hospitals, street lighting, public buildings	Creation and financing of utility-based ESCO for projects in public and private sectors	Yes
Serbia—Energy efficiency project	2004	Schools and hospitals	The finance installation of retrofits in 17 schools and 12 hospitals, mostly for heating-related improvements	No
Uruguay—Energy efficiency	2004	Public lighting	Utility demand-side management (DSM) component targets EE equipment financing through utility bills for government sector	No
Burkina Faso—Power sector development	2004	Public administration buildings	Support of utility DSM education and investments, beginning with piloting approaches in public administration buildings before expanding the program to industrial, commercial, and residential customers	No
China—Heat reform and building energy efficiency	2005	Residential buildings	Demonstration projects to upgrade heating-related performance	No
Armenia—Urban heating	2005	Municipalities and schools	Financing EE upgrades in municipalities and schools to improve heating services	No
Armenia—Yerevan water and wastewater	2005	Municipal water utilities	The financing of investments to reduce energy use in municipal water pumping	No
Macedonia—Sustainable energy	2006	Publicly owned buildings, schools	Creation of a utility-based ESCO and demonstrating ESPC concept in public buildings; parallel guarantee facility could also cover late payments to ESCO from public customers	Yes
Belarus—Post-Chernobyl recovery	2006	Public buildings	The upgrading of heat production and distribution equipment and improved thermal insulation and lighting in public buildings	No

(continued)

Project	Year approved	Public sector targeted	Specific components	Use of ESPCs
Ukraine—Urban infrastructure	2007	Municipal water utilities	Investments in EE for municipal water/wastewater utilities	No
Russia—Housing and communal services	2008	Housing and communal services	Upgrading and retrofitting existing housing and communal service infrastructure, including heating and water supply, sewerage, housing energy supply, and housing	No
Moldova—National water supply and sanitation	2008	Municipal water supply	Rehabilitation of water treatment plant and pumping stations, energy audits, hydrological optimizations, retrofit of electromechanical equipment, and exploration of sector-wide CDM potential	No
Moldova—Additional financing for energy	2009	Hospitals, schools, kindergartens, and social service centers	The financing of heating retrofitting of public buildings	No
Montenegro—Energy efficiency	2009	Public buildings	The financing of EE improvement of heating system and building envelope (among others) of public sector buildings	No
International Finance Corporation – Global Environment Facility				
ELI—Tranche 2 (Czech Republic, Hungary, Latvia, Philippines)	2000	Street lighting	Technical assistance and transaction support for efficient lighting programs	No
Hungary Energy Efficiency Cofinancing Program (Tranche 2)	2001	Street lighting, schools	Loan guarantee program to support commercial EE projects	Yes

Commercializing Energy Efficiency Financing (CEEF)—Tranches 1 & 2 (Czech Republic, Estonia, Latvia, Lithuania, Slovak Republic)	2002 (phase 1) 2005 (phase 2)	Municipal buildings, street lighting, schools	Development of financing program for EE	Yes
World Bank—Carbon Finance				
India—Karnataka municipal water supply energy efficiency	In preparation	Municipal water supply	Sales of emissions from EE investments in municipal water supply systems in six municipalities in Karnataka	No
India—Energy efficient street lighting project	In preparation	Street lighting	Sales of emissions from EE investments in up to seven municipal street lighting systems	Yes

Note: Projects include energy efficiency components that explicitly target public sector facilities and services but do not include supply-side energy efficiency improvements in power and heating utilities.

Sample of Non–World Bank Projects with Public Sector Energy Efficiency Components (2000–09)

Project	Country	Donor	Year	Specific component	Use of ESPCs?
Water and Energy Savings ("Watergy")	Brazil, India, Mexico, South Africa	USAID	1997	Alliance to Save Energy program to provide capacity building and technical assistance (TA) to water utilities to promote water and energy efficiency (EE)	Yes
Municipal Energy Efficiency Program (MEEP)	Bulgaria	USAID	2000	TA and partial loan guarantee for municipal EE investments	No
Promoting an Energy-Efficient Public Sector (PEPS)	China, India, Mexico	USAID, US EPA	2001	Lawrence Berkeley National Laboratory program to provide TA to national/provincial/municipal governments to adopt energy-efficient procurement of office and other equipment	No
Public Sector Energy Efficiency Program	Hungary	UNDP/GEF	2001	Promotion for implementation of EE projects in municipalities, hospitals, and other public institutions	Yes
Municipal Network for Energy Efficiency (MUNEE)	Eastern Europe Regional	USAID	2001	Alliance to Save Energy program to provide TA to municipalities to implement EE improvements in heating/water systems and buildings, particularly schools and hospitals	No
Removing Barriers to Green House Gas (GHG) Emissions Mitigation through Energy Efficiency in the District Heating (DH) System - Ph 1 & 2	Ukraine	UNDP/GEF	2001 (Ph 1) 2006 (Ph 2)	Creation of publicly owned ESCO (Rivne City) to provide risk sharing and financing of upgrades for the municipal heating system and its customers	Yes
Cost-Effective Energy Efficiency Measures in the Education Sector	Russia	UNDP/GEF	2002	TA and creation of revolving fund to support EE retrofits in schools	Yes

Project	Country	Funder	Year	Description	
Demand-Side Energy Efficiency in Public Buildings, Lodz Municipal ESCO	Poland	EBRD/GEF	2003	Creation of an innovative municipal ESCO to support bundling of municipal buildings for EE retrofits	Yes
Energy Conservation II	India	USAID	2005	Development of a framework for a state energy conservation fund in Maharashtra that would finance public and private EE projects	No
Public Sector Energy Efficiency Procurement	Egypt, Mexico	USAID	2005	Development of bidding documents and investment grants to support ESPCs in public sector	Yes
Removing Barriers to the Reconstruction of Public Lighting	Slovak Republic	UNDP/GEF	2005	Support retrofits of municipal street lighting schemes with special fund and creation of public ESCO as transaction broker	Yes
Energy Efficiency Public Lighting	Vietnam	UNDP/GEF	2005	TA to support efficient lighting in street lighting, schools and other public facilities	No
ESCO Financing 2	Ukraine	EBRD	2005	Development of mechanisms for implementing EE projects in public sector and small and medium enterprises using private financing and public ESCO (UkrESCO)	Yes
Building the Local Capacity for Promoting Energy Efficiency in Private and Public Buildings	Bulgaria	UNDP/GEF	2006	Promotion and TA for improving EE in public buildings through code enhancement, ESCOs training and awareness raising	Yes
Energy Conservation III	India	USAID	2007	Municipal tender development for bundled municipal EE projects in Gujarat using EPCs	Yes
ESCO Fund	Bulgaria	EBRD	2007	Creation of loan fund to purchase receivables under ESCO contracts, which will include municipal EE projects	Yes

(continued)

Project	Country	Donor	Year	Specific component	Use of ESPCs?
Cherkasy Energy Efficiency	Ukraine	EBRD	2007	Financing of EE investments in residential buildings	No
Odessa District Heating	Ukraine	EBRD	2008	Financing of EE investments in residential buildings	No
ECO-Asia	South and East Asia Regional	USAID, ADB	2008	TA to establish state energy conservation funds in India (Kerala and Madhya Pradesh)	TBD
Improving Efficiency in Public Buildings	Russia	EBRD/GEF	2008	Creation of financing facility to purchase receivables under ESCO contracts, which will include municipal EE projects	Yes
Energy Efficiency for Caribbean Water and Sanitation Companies	Caribbean Regional	IADB	2008	TA to develop EE investment/action plans for regional water and sanitation utilities	No
Energy Efficiency	Philippines	ADB	2009	Promotion and financing of EE projects to reduce public energy expenditures and develop sustainable ESCO business models, including creation of public ESCO	Yes
Petropavlovsk-Kamchatsky Municipal Water	Russia	EBRD	Under preparation	Rehabilitation and upgrade of water pumping stations with the objective to improve EE of operations	TBD
Improving Urban Housing Efficiency	Russia	EBRD/GEF	Under preparation	Integration of EE into housing planning, refurbishment, and maintenance through a municipal housing reform fund and building code	No

Note: IADB = Inter-American Development Bank.

Index

Boxes, figures, notes, and tables are indicated by *b*, *f*, *n*, and *t*, respectively.

www.ingramcontent.com/pod-product-compliance
Lightning Source LLC
Chambersburg PA
CBHW061151220326
41599CB00025B/4436